Lecture Notes in Mathematics

Edited by A. Dold and B. Eckmann

723

Willy Brandal

Commutative Rings whose Finitely Generated Modules Decompose

Springer-Verlag
Berlin Heidelberg New York 1979

Author

Willy Brandal
Department of Mathematics
University of Tennessee
Knoxville, TN 37916/USA

AMS Subject Classifications (1970): 13-02, 13C05, 13F05, 13F10, 13G05

ISBN 3-540-09507-1 Springer-Verlag Berlin Heidelberg New York
ISBN 0-387-09507-1 Springer-Verlag New York Heidelberg Berlin

Library of Congress Cataloging in Publication Data. Brandal, Willy, 1942- The commutative rings whose finitely generated modules decompose. (Lecture notes in mathematics ; v. 723) Bibliography: p. Includes index. 1. Commutative rings. 2. Modules (Algebra) 3. Decomposition (Mathematics) I. Title. II. Series: Lecture notes in mathematics (Berlin) ; v. 723. QA3.L28 no. 723 [QA251.3] 510'.8s [512'.4] 79-15959

Printing and binding: Beltz Offsetdruck, Hemsbach/Bergstr.
2141/3140-543210

Table of Contents

Introduction

Throughout all rings will be commutative with identities, all modules will be unitary modules, and R will always denote a ring. R is said to be an FGC ring if every finitely generated R-module decomposes into a direct sum of cyclic submodules.

The purpose of these notes is to describe all the FGC rings; i.e., characterize the FGC rings and give as many examples as possible. One form of the Fundamental Theorem of Abelian Groups says that the ring of integers is an FGC ring. Another form of this theorem says that P.I.D.'s are FGC rings. Thus we present a generalization of the Fundamental Theorem of Abelian Groups.

The level of exposition is such that these notes should be readable to as large an audience as possible. The only background necessary is a good first year graduate course in algebra and point set topology. The algebra background of the reader should include field theory and the following concepts from commutative algebra: prime ideals, modules, free modules, localizations, and tensor products. To make these notes accessible to as large an audience as possible, no homological algebra is included. For example "injective modules" and "projective modules" are never discussed, and there are no exact sequences. With this background, these notes are self-contained in the sense that the reader is never refered to another source to complete an argument that is essential in the main development. The use of references is only for historical purposes or to point out related topics which are not needed in the main development.

A number of results were originally done using more advanced techniques. For example the discussion of h-local domains in section 2 was originally done by E. Matlis using homological algebra. Our proof of the main properties of h-local domains (2.6) is therefore new, more elementary, and as indicated lends itself to generalizations. As another example, some authors prove the uniqueness of decomposition of finitely generated modules over valuation rings (3.4) using exterior algebras. The proof presented here is one using determinants, so that the reader need not be familiar with exterior algebras. In each such case it is our own belief that the arguments presented here are efficient, i.e., are as short or almost as short as arguments using advanced techniques, especially if these advanced concepts were to be explained.

We briefly discuss the recent developments of the main theorem in chronological order. The early interest in FGC rings centered around domains. As late as 1950 the only known FGC domains were the P.I.D.'s. In 1952 I. Kaplansky showed that almost maximal valuation domains were FGC domains. For another 20 years the only known FGC domains were the P.I.D.'s and the almost maximal valuation domains.

In the mid 1960's E. Matlis, while working on another problem, studied h-local domains which are describable in terms of the decomposition of torsion modules. In 1967 R.S. Pierce characterized the commutative regular FGC rings as finite products of fields. Of significance was his relating the decomposition of modules over a ring to topological considerations of the spectrum of that ring.

In 1973 W. Brandal, and independently in 1974 T. Shores and R. Wiegand, showed that almost maximal Bezout domains are FGC domains, and that there exists an example due to B. Osofsky of an almost maximal Bezout domain which is not a P.I.D. and not a valuation domain. T. Shores and R. Wiegand in this same paper, showed that torch rings are FGC rings and the topological discussions begun by R.S. Pierce were greatly improved. Thus in 1974 the known FGC rings included all rings which are finite products of rings of the following three types: (almost) maximal valuation rings, almost maximal Bezout domains, and torch rings.

For the main theorem characterizing the FGC rings it remained to be shown that these were the only FGC rings. In 1975 S. Wiegand showed that in an FGC domain every non-zero prime ideal is a subset of only one maximal ideal. In 1976 W. Brandal and R. Wiegand showed that an FGC rings can have only finitely many minimal prime ideals, and hence the reduced (no non-zero nilpotent elements)

FGC rings were characterized. As a special case the FGC domains were characterized as the almost maximal Bezout domains. P. Vamos, and independently T. Shores, gave a generalization of the S. Weigand result mentioned above. These facts were combined so that before the end of 1976 a characterization of the FGC rings were presented in an expository article by R. Wiegand and S. Wiegand.

These notes are not given in the chronological order as described in the last few paragraphs. There are two parts. Part one consists of nine sections where the main theorem characterizing the FGC rings is given. Part two consists of six sections where examples of FGC rings are presented.

Section one introduces linearly compact modules, maximal rings, and almost maximal rings. The main fact is that a maximal ring is a finite product of local maximal rings.

Section two introduces h-local domains. A domain R is h-local if and only if every torsion R-module T is a direct sum of the localizations T_M where M ranges over all the maximal ideals of R. A domain is an almost maximal ring if and only if it is h-local and locally almost maximal.

Section three introduces valuation rings and Bezout rings. A valuation ring which is not a domain is almost maximal if and only if it is maximal. If a module over a valuation decomposes into a direct sum of cyclics, then there is a uniqueness of decomposition. Bezout rings are Prufer rings.

Section four gives some basic facts about FGC rings and treats the local case. Namely, a local ring is an FGC ring if and only if it is an almost maximal valuation ring. An FGC ring is a locally almost maximal Bezout ring.

Section five gives further results about FGC rings and introduces torch rings. Almost maximal Bezout domains are FGC rings. Torch rings are FGC rings. If R is an FGC ring with a unique minimal prime ideal P, then P is a uniserial R-module. In an FGC domain, every non-zero prime ideal is a subset of a unique maximal ideal.

The next major fact to be proved is that an FGC ring has only finitely many minimal prime ideals. This requires the topological considerations of the next three sections.

Section six introduces the Zariski and patch topologies of the spectrum of a ring. The patch topology makes specR a Boolean space. If Y is a patch of specR, the Zariski and patch subspace topologies of minY are the same.

Section seven introduces the Stone-Cech compactification of N, βN, where N is the set of natural numbers. More generally, if X is a non-empty discrete topological space and βX denotes the set of all ultrafilters of X, then βX is a Stone-Cech compactification of X. If X is a non-empty discrete topological space and C is an infinite closed subset of βX, then C contains a closed subset homeomorphic to βN and a closed subset homeomorphic to $\beta N - N$. There exists a c-point in $\beta N - N$.

Section eight relates topology to the decomposition of modules. If specR contains three pairwise disjoint Zariski open subsets whose patch closures intersect, then R is not an FGC ring. If specR contains a thin patch with a 3-point relative to the patch topology, then R is not an FGC ring. A Boolean space with a countably infinite dense subset consisting of isolated points either has a 3-point or has a subspace homeomorphic to $\beta N - N$. The topological considerations of the last three sections are combined to show that an FGC ring has only finitely many minimal prime ideals.

Section nine presents the main theorem characterizing the FGC rings. R is an FGC ring if and only if R is a finite product of rings of the following three types: maximal valuation rings, almost maximal Bezout domains, and torch rings. An alternate characterization of FGC rings is presented. If R is an FGC ring and A is a finitely generated R-module, then there is a uniqueness of decomposition of A into indecomposable cyclics and there is a uniqueness of canonical form decomposition of A. As a special case of the main theorem, a domain is an FGC domain if and only if it is an almost maximal Bezout domain. If R is an FGC domain and A is a finitely generated R-module, then there is a uniqueness of decomposition of A into a direct sum of a free submodule and a sum of primary torsion submodules. This completes part one.

Part two gives the construction of examples. The first three sections are preliminaries and the last three sections consist of the examples of the three types of indecomposable FGC rings listed in the main theorem. Section ten introduces valuations, giving several of the standard results about valuations. For example, a valuation over a field can be extended to a valuation on any field extension.

Section eleven introduces long power series rings. Given any field K and totally ordered group G there is a long power series ring R which is a maximal valuation domain, K is the residue field of R , and G is the divisibility group of R .

Section twelve introduces maximally complete valuation domains. A valuation domain is maximal if and only if it is maximally complete. As a consequence, if K is an algebraically closed field and G is a finite direct sum of copies of the rationals with the lexicographic ordering, and R is the long power series ring relative to K and G , then the quotient field of R is an algebraically closed field.

Section thirteen briefly gives a list of maximal valuation rings. Except perhaps for the long power series rings, these are well known.

Section fourteen gives examples of almost maximal Bezout domains. P.I.D.'s and almost maximal valuations domains are well known examples of almost maximal Bezout domains. The first almost maximal Bezout domain that is not a P.I.D. nor a valuation ring is the example of B. Osofsky. It has exactly two maximal ideals and the localization at each maximal ideal is a maximal valuation domain of Krull dimension one. A generalization is given, namely an example of S. Wiegand. If X is a finite tree with zero such that the intersection of any two distinct maximal chains of X is {0} , then there exists an almost maximal

Bezout domain R such that X is order isomorphic to specR and R_P is a maximal valuation domain for all prime ideals P of R . Another generalization is given, namely an example of P. Vamos. There exists an almost maximal Bezout domain R with countably infinite many maximal ideals such that R_M is a Krull dimension one maximal valuation domain for all maximal ideals M of R .

Section fifteen gives the construction of torch rings using the three examples of the last paragraph.

A preliminary version of much of part one was presented in a seminar the first half of the school year 1976-77 at the University of Tennessee. The participants were D. Anderson, J.H. Carruth, J. Cunningham, D. Dobbs, E. Evans, W. Keigher, R. McConnel, and R. Rowlett. I want to express my appreciation for many helpful comments by these participants.

Before beginning the text, we introduce some notations to be used throughout these notes. As already mentioned R denotes a ring, where ring means commutative ring with identity. $\underline{N}$ will denote the set of natural numbers {1,2,3,...} . When topological considerations are made, N will be assumed to have the discrete topology. $\underline{Z}$ will denote the set of integers, and may be considered as a ring, as an additive Abelian group, or as a totally ordered Abelian group with the standard ordering. <u>specR</u> will denote the set of all prime ideals of R , and <u>mspecR</u> will denote the set of all maximal ideals of R . If X is a set, then $|X|$ will denote the cardinality of X .

PART 1 PROVING THE MAIN THEOREM

Section 1 Linearly Compact Modules and Almost Maximal Rings

The material in this section is due to D. Zelinsky [38]. He studied linear compactness for topological modules, where in that definition the cosets must be closed. Ignoring the topology as we have done, amounts to assuming that all the topologies are discrete. Other papers entitled "Linearly compact modules and rings" have been written by H. Leptin [19], [20], and S. Warner [33].

A family of sets is said to have the <u>finite intersection property</u>, abbreviated <u>f.i.p.</u> , if the intersection of every finite subfamily is non-empty.

<u>Definition</u>: Let A be an R-module. A is a <u>linearly compact</u> R-module if whenever $\{x_\alpha + A_\alpha\}_{\alpha \in X}$ is a family of cosets of submodules of A ($x_\alpha \in A$ and A_α is a submodule of A) with the f.i.p., then $\bigcap_{\alpha \in X} x_\alpha + A_\alpha \neq \emptyset$.

One can translate this into a condition about solving congruences. With the above notation $x \in x_\alpha + A_\alpha$ if and only if $x \equiv x_\alpha \bmod A_\alpha$. Thus an R-module A is linearly compact if given any family of congruences $\{x \equiv x_\alpha \bmod A_\alpha\}_{\alpha \in X}$ of A , being able to find a solution for any finite subset of these congruences implies one can find a solution for all the congruences.

<u>Lemma 1.1</u>: Let A be an R-module, $z, x_1, x_2 \in A$, and A_1 and A_2 submodules of A . Then $z \in (x_1 + A_1) \cap (x_2 + A_2)$ if and only if $(x_1 + A_1) \cap (x_2 + A_2) = z + (A_1 \cap A_2)$.

<u>Proof:</u> Trivial.

<u>Proposition 1.2:</u>

1. A submodule of a linearly compact R-module is linearly compact.
2. A homomorphic image of a linearly compact R-module is linearly compact.
3. If B is a submodule of the R-module A with both B and A/B being linearly compact, then A is linearly compact.
4. A finite direct sum of linearly compact R-modules is linearly compact.

5. An R-module with the descending chain condition on submodules is linearly compact.

Proof:

1. Trivial.

2. Suppose $f: A \longrightarrow B$ is an R-epimorphism and A is a linearly compact R-module. Let $\{x_\alpha + B_\alpha\}_{\alpha \in X}$ be a family of cosets of submodules of B with the f.i.p. . For each $\alpha \in X$ choose $y_\alpha \in f^{-1}(x_\alpha)$. $\{y_\alpha + f^{-1}(B_\alpha)\}_{\alpha \in X}$ has the f.i.p. . A is linearly compact implies there exists $z \in \bigcap_{\alpha \in X} y_\alpha + f^{-1}(B_\alpha)$, and so $f(z) \in \bigcap_{\alpha \in X} x_\alpha + B_\alpha$, showing B is linearly compact.

3. Let $\{x_\alpha + A_\alpha\}_{\alpha \in X}$ be a family of cosets of submodules of A with the f.i.p., and let $p: A \longrightarrow A/B$ be the canonical projection. Using 1.1 we may assume that this family is closed under finite intersections. $\{p(x_\alpha) + p(A_\alpha)\}_{\alpha \in X}$ has the f.i.p. . Since A/B is linearly compact, there exists $z \in \bigcap_{\alpha \in X} p(x_\alpha) + p(A_\alpha)$. Then $z = y + B$ for some $y \in A$. $B \cap (x_\alpha - y + A_\alpha) \neq \emptyset$ for all $\alpha \in X$. Since $\{x_\alpha + A_\alpha\}_{\alpha \in X}$ is closed under finite intersections, $\{B \cap (x_\alpha - y + A_\alpha)\}_{\alpha \in X}$ is closed under finite intersections and so has the f.i.p. . B is linearly compact implies there exists $x \in \bigcap_{\alpha \in X} (B \cap (x_\alpha - y + A_\alpha))$. Thus $x + y \in \bigcap_{\alpha \in X} x_\alpha + A_\alpha$, showing that A is linearly compact.

4. This follows from part 3 and induction.

5. This follows easily using 1.1 . q.e.d.

Example 1.3: Let A be an R-module such that A has as a submodule $\bigoplus_{n \in N} A_n$ where $A_n \neq \{0\}$ for all $n \in N$. Then A is not linearly compact.

Proof: For $n \in N$ choose $y_n \in A_n - \{0\}$, let $B_n = \sum_{i>n} A_i$, and let $x_n = \sum_{i=1}^{n} y_i$

Then $\mathcal{A} = \{\bigoplus_{n\in N} A_n\} \cup \{x_n + B_n\}_{n\in N}$ is a family of cosets of submodules of A with the f.i.p., yet $\cap\mathcal{A} = \emptyset$, showing A is not linearly compact. q.e.d.

Definition: R is a maximal ring if R is a linearly compact R-module. R is an almost maximal ring if R/I is a linearly compact R-module for all non-zero ideals I of R.

The use of the term "maximal" will be discussed in section 12. If I is a proper ideal of R, then R/I is a linearly compact R-module if and only if R/I is a linearly compact R/I-module. If R is a maximal ring, then by 1.2(2), R is an almost maximal ring. Note also that R is an almost maximal ring if and only if whenever $\{r_\alpha + I_\alpha\}_{\alpha\in X}$ is a family of cosets of ideals of R with the f.i.p. such that $\bigcap_{\alpha\in X} I_\alpha \neq \{0\}$, then $\bigcap_{\alpha\in X} r_\alpha + I_\alpha \neq \emptyset$.

Consider the ring of integers, Z. Z is an almost maximal ring, but Z not a maximal ring. For if I is a non-zero ideal of Z, then Z/I satisfies the descending chain condition and so is linearly compact by 1.2(5). This verifies that Z is an almost maximal ring. Let $p_1, p_2, \ldots$ be a listing of the odd prime integers. Then $\mathcal{A} = \{1 + 2Z\} \cup \{p_n Z\}_{n\in N}$ is a family of cosets of ideals of Z with the f.i.p. and $\cap\mathcal{A} = \emptyset$. This verifies that Z is not a maximal ring. Similarly one can verify that a P.I.D. is an almost maximal ring.

Recall that for a ring R, $\cap\text{mspec}R$ is the Jacobson radical of R. We say R is local if $\text{mspec}R$ has only one element (no Noetherian condition is included), and R is semilocal if $\text{mspec}R$ is a finite set.

Lemma 1.4: If R is a maximal ring, then R is semilocal.

Proof: Assume R is a maximal ring. Suppose for each $M \in \text{mspec}R$ one has $y_M \in R$. By the Chinese Remainder theorem $\{y_M + M\}_{M\in\text{mspec}R}$ has the f.i.p.. Since R is maximal, $\bigcap_{M\in\text{mspec}R} y_M + M \neq \emptyset$.

Define $I = \{r \in R\colon\ r \in M$ for all but a finite number of $M \in \text{mspec}R\}$.

We claim $\{I\} \cup \{1 + M\}_{M \in \text{mspec}R}$ has the f.i.p.. For if one has the subfamily $\{I, 1 + M_1, \ldots, 1 + M_n\}$ with $\{M_1, \ldots, M_n\} \subset \text{mspec}R$, then $(\bigcap_{i=1}^{n} 1 + M_i) \cap I \supset (\bigcap_{i=1}^{n} 1 + M_i) \cap (\bigcap (\text{mspec}R - \{M_1, \ldots, M_n\})) \neq \emptyset$ by the first paragraph. Thus $\{I\} \cup \{1 + M\}_{M \in \text{mspec}R}$ has the f.i.p., and since R is a maximal ring there exists $x \in I \cap (\bigcap_{M \in \text{mspec}R} 1 + M)$. Then $x \notin M$ for all $M \in \text{mspec}R$. But $x \in I$ then implies mspecR is finite. <u>q.e.d.</u>

1.4. gives an alternate proof that Z is not a maximal ring. The following is refered to as a "lifting idempotents" result.

<u>Lemma 1.5</u>: If R is a maximal ring and f is a idempotent of $R/\cap\text{mspec}R$, then there exists an idempotent e of R such that $e + \cap\,\text{mspec}R = f$.

<u>Proof</u>: Define $\mathcal{A} = \{d + I: I$ is an ideal of R, $I \subset \cap\text{mspec}R$, $d \in R$, $d + \cap\,\text{mspec} = f$, and $d^2 - d \in I\}$. Order $\mathcal{A}$ by $d + I \geq d' + I'$ if $d + I \subset d' + I'$. This is a partial ordering of $\mathcal{A}$, and $f \in \mathcal{A}$ so $\mathcal{A} \neq \emptyset$.

Suppose $\mathcal{C} = \{e_\alpha + I_\alpha\}_{\alpha \in X}$ is a non-empty chain of $\mathcal{A}$. Then $\mathcal{C}$ has the f.i.p.. Since R is a maximal ring, there exists $x \in \cap\mathcal{C}$. We claim that $x + \bigcap_{\alpha \in X} I_\alpha \in \mathcal{A}$. For $\alpha \in X$, $x = e_\alpha + r_\alpha$ for some $r_\alpha \in I_\alpha$, and so $x^2 - x = (e_\alpha + r_\alpha)^2 - (e_\alpha + r_\alpha) = e_\alpha{}^2 - e_\alpha + r_\alpha(2e_\alpha + r_\alpha - 1) \in I_\alpha$. Thus $x^2 - x \in \bigcap_{\alpha \in X} I_\alpha$, and it follows that $x + \bigcap_{\alpha \in X} I_\alpha \in \mathcal{A}$ and thus is an upper bound for $\mathcal{C}$. By Zorn's Lemma, there exists a maximal element $e + J$ in $\mathcal{A}$.

We wish to show that $e^2 = e$. $(1 - 2e)^2 = 1 + 4(e^2 - e)$ and $e^2 - e \in J \subset \cap\text{mspec}R$ implies $1 - 2e$ is a unit of R. Let $y = (e^2 - e)(1 - 2e)^{-1}$. Then $y \in J$ and one checks that $(e + y)^2 - (e + y) = y^2$. Define $d = e + y$ and $I = Ry^2$. Then $d + I \in \mathcal{A}$ and $d + I \geq e + J$. By the maximality of $e + J$, $d + I = e + J$. In particular $J = I = Ry^2$. But $y \in J = Ry^2$, so $y = ry^2$ for some $r \in R$. $y(1 - ry) = 0$ and $y \in J \subset \cap\,\text{mspec}\,R$ implies $1 - ry$ is a unit of R, and so $y = 0$. The identity $(e + y)^2 - (e + y) = y^2$ and

$y = 0$ gives $e^2 = e$. Clearly $e + J \in \mathcal{A}$ implies $e + \cap\, \mathrm{mspec}R = f$. q.e.d.

1.5 is not true if the condition "maximal" is replaced by "almost maximal." For example, let $R = Z_{Z-(2Z \cup 3Z)}$. Then R is almost maximal, $R/\cap\, \mathrm{mspec}R \cong Z/2Z \oplus Z/3Z$, and idempotents there cannot be lifted to idempotents in R. In 1.5 one could show that the element e is unique, although this will not be needed in our development.

Theorem 1.6: (D. Zelinsky [38]) If R is a maximal ring, then R is a finite direct product of local rings, each of which is clearly a maximal ring.

Proof: By 1.4 R is semilocal. Let $\mathrm{mspec}R = \{M_1,\dots,M_n\}$. By the Chinese Remainder Theorem $R/\cap\mathrm{mspec}R = R/\bigcap_{i=1}^{n} M_i \cong \prod_{i=1}^{n} R/M_i$. Hence

$R/\cap\, \mathrm{mspec}R = \prod_{i=1}^{n} (R/\cap\, \mathrm{mspec}\ R)f_i$ where the f_i's are orthogonal idempotents of $R/\cap\, \mathrm{mspec}R$. By 1.5 there exist idempotents e_i of R such that $e_i + \cap\mathrm{mspec}R = f_i$. For $i \neq j$, $e_ie_j \in \cap\mathrm{mspec}R$ and $(e_ie_j)^2 = e_ie_j$, and so $e_ie_j = 0$. This says that the e_i's are also orthogonal idempotents. For each j, $(1 - \sum_{i=1}^{n} e_i)\, e_j = e_j - \sum_{i=1}^{n} e_ie_j = e_j - e_j^2 = 0$, and so $1 - \sum_{i=1}^{n} e_i \in \mathrm{Ann}_R(e_j)$. It follows that $R = (\prod_{i=1}^{n} Re_i) \oplus (\bigcap_{j=1}^{n} \mathrm{Ann}_R(e_j))$. $1 = e_1 + \dots + e_n + a$ for some $a \in \bigcap_{j=1}^{n} \mathrm{Ann}_R(e_j)$. $a = 1 \cdot a = (e_1 + \dots e_n + a)a = a^2$, $a = a^2$, and $a \in \cap\, \mathrm{mspec}R$ implies $a = 0$. Thus $R = \prod_{i=1}^{n} Re_i$. Each Re_i must be a local ring, for otherwise mspecR would have more than n elements. q.e.d.

Corollary 1.7: If R is a maximal domain, then R is local.

Proof: Trivial.

Section 2 h-local Domains

We begin this section with some preliminary localization results. For $P \in \mathrm{spec}R$ one has the canonical ring homomorphism $\pi_P\colon R \to R_P$ given by $\pi_P(r) = r/1$ for $r \in R$. If also I is an ideal of R, then $I_P = \pi_P(I)_P$. For an ideal J of R_P we use $\underline{J^c}$ to denote $\pi_P^{-1}(J)$, and of course J^c is an ideal of R.

<u>Lemma 2.1</u>: If I is an ideal of R, then $I = \bigcap_{M \in \mathrm{mspec}R} (I_M)^c$.

<u>Proof:</u> The one inclusion is clear. On the other hand let $x \in \bigcap_{M \in \mathrm{mspec}R} (I_M)^c$ and define $J = \{r \in R\colon rx \in I\}$. Suppose $M \in \mathrm{mspec}R$. $x \in I_M^c$ implies $x/1 = \pi_M(x) = i/s$ for some $s \in R - M$ and $i \in I$. There exists $t \in R - M$ such that $(xs - i)t = 0$. Hence $st \in J \cap (R - M)$ and so $J \not\subset M$. M is an arbitrary element of $\mathrm{mspec}R$, so $J = R$, $1 \in J$, and $x \in I$. The desired equality holds.

<u>q.e.d.</u>

<u>Corollary 2.2</u>:

1. Let I and J be ideals of R. Then $I \supset J$ if and only if $I_M \supset J_M$ for all $M \in \mathrm{mspec}R$.
2. Let I and J be ideals of R. Then $R/I \cong R/J$ if and only if $(R/I)_M \cong (R/J)_M$ for all $M \in \mathrm{mspec}R$.
3. Let R be a domain with quotient field Q. If A is an R-submodule of Q, then $A = \bigcap_{M \in \mathrm{mspec}R} A_M$, where all the sets are considered as subsets of Q.

<u>Proof</u>:

1. If $I_M \supset J_M$ for all $M \in \mathrm{mspec}R$, then $I = \bigcap_{M \in \mathrm{mspec}R} (I_M)^c \supset \bigcap_{M \in \mathrm{mspec}R} (J_M)^c = J$. The converse is trivial.
2. $R/I \cong R/J$ if and only if $I = J$. $(R/I)_M \cong (R/J)_M$ if and only if $I_M = J_M$. Thus this follows from part 1.
3. If $A \subset R$, then this is a special case of 2.1. If $A \not\subset R$, then a simila

argument as the proof of 2.1 can be given. <u>q.e.d.</u>

The following well known result is included for the sake of easy reference.

<u>Lemma 2.3</u>: Let S be a multiplicatively closed subset of R and let A be an R-module.

1. If B and C are submodules of A, then $(B \cap C)_S = B_S \cap C_S$.
2. If $\{A_\alpha\}_{\alpha\in X}$ is a family of submodules of A, then $(\sum_{\alpha\in X} A_\alpha)_S = \sum_{\alpha\in X} (A_\alpha)_S$.

<u>Proof</u>: Straight forward.

If X is a set, then $\{X_\alpha\}_{\alpha\in Y}$ is a <u>partition</u> of X if $X = \bigcup_{\alpha\in Y} X_\alpha$ and $X_\alpha \cap X_\beta = \emptyset$ if $\alpha \neq \beta$. Such a partition is <u>nontrivial</u> if $X_\alpha \neq \emptyset$ for all $\alpha \in Y$. If I is an ideal of R, then we use the notation <u>mspec (I)</u> = $\{M \in \text{mspec} R : I \subset M\}$.

<u>Lemma 2.4</u>: Let I be an ideal of R. If mspec (I) is finite, then R/I is a direct sum of indecomposable R-modules.

<u>Proof</u>: If $R/I \cong \bigoplus_{i=1}^{n} R/I_i$ for ideals I_i of R, then $\{\text{mspec}(I_i): i=1,..,n\}$ is a partition of mspec (I). If one decomposes R/I, R/I_i, etc., then mspec (I) is partitioned into smaller sets. Since mspec (I) is finite, this process must stop after a finite number of steps, at which time R/I is a direct sum of indecomposable R-modules. <u>q.e.d.</u>

The following result was done by W. Brandal [3] in the special case where R is a Prufer domain.

<u>Proposition 2.5</u>: Let I be an ideal of R with mspec (I) finite. Then R/I is indecomposable if and only if for all nontrivial partition $\{\mathcal{M}_1, \mathcal{M}_2\}$ of mspec (I) there exists $M_1 \in \mathcal{M}_1$, $M_2 \in \mathcal{M}_2$ and $P \in \text{spec} R$ such that $I \subset P \subset M_1 \cap M_2$.

<u>Proof</u>: Suppose we have the condition about partitions and suppose R/I is decomposable. Then $R/I \cong R/I_1 \oplus R/I_2$ where I_1 and I_2 are proper ideals of R. $\{\text{mspec}(I_1), \text{mspec}(I_2)\}$ is a nontrivial partition of mspec (I), so by assump-

tion there exist $M_1 \in \text{mspec}(I_1)$, $M_2 \in \text{mspec}(I_2)$, and $P \in \text{spec}R$ such that $I \subset P \subset M_1 \cap M_2$. $I_1 \cap I_2 = I \subset P$ implies $I_1 \subset P$ or $I_2 \subset P$. Without loss of generality assume $I_1 \subset P$. Then $M_2 \in \text{mspec}(I_1) \cap \text{mspec}(I_2)$, a contradiction.

Conversely, suppose R/I is indecomposable. Let $\{\mathcal{M}_1, \mathcal{M}_2\}$ be a nontrivial partition of mspec (I). By assumption mspec (I) is finite, and so can be labeled so as to have $\mathcal{M}_1 = \{P_1,\ldots,P_k\}$ and $\mathcal{M}_2 = \{P_{k+1},\ldots,P_n\}$ where $1 \leq k < n$.

We claim that there exist $r \in \{1,\ldots,k\}$ and $s \in \{k+1,\ldots,n\}$ such that ($x \in R - P_s$ implies $R_{P_r} x \not\subset I_{P_r}$) or ($y \in R - P_r$ implies $R_{P_s} y \not\subset I_{P_s}$) . To verify this claim, we suppose it is false and derive a contradiction. Thus for each $i = 1, \ldots, k$ and $j = k+1, \ldots, n$ there exists $x_{ij} \in R - P_j$ with $R_{P_i} x_{ij} \subset I_{P_i}$ and there exists $y_{ij} \in R - P_i$ with $R_{P_j} y_{ij} \subset I_{P_j}$. Define $L' = \bigcap_{i=1}^{k} (\sum_{j=k+1}^{n} Rx_{ij})$ and $L'' = \bigcap_{j=k+1}^{n} (\sum_{i=1}^{k} Ry_{ij})$. Using 2.3 one sees that $L'_{P_i} \subset I_{P_i}$, $L' \not\subset P_j$, $L'' \not\subset P_i$, and $L''_{P_j} \subset I_{P_j}$ for all $i = 1, \ldots, k$ and for all $j = k+1, \ldots, n$. By localizing (using 2.2(1) and 2.3) $(I+L') \cap (I + L'') = I$. $(I + L') + (I + L'') = R$ since the left hand side is not a subset of any maximal ideal of R . By the Chinese Remainder Theorem $R/I \cong R/(I + L'') \oplus R/(I + L'')$, contradicting R/I is indecomposable. This verifies the claim and so without loss of generality we may assume that there exists $r \in \{1,\ldots,k\}$ and $s \in \{k+1,\ldots,n\}$ such that $x \in R - P_s$ implies $R_{P_r} x \not\subset I_{P_r}$.

Let $\mathcal{J} = \{J:$ J is an ideal of R_{P_r} , $I_{P_r} \subset J$, and $J^c \cap (R - P_s) = \emptyset\}$. By the last paragraph $I_{P_r} \in \mathcal{J}$ and so $\mathcal{J} \neq \emptyset$. By Zorn's Lemma $\mathcal{J}$ has a maximal element J_0 , and it follows that J_0 is a prime ideal of R_{P_r} .

Let $\mathcal{I} = \{L:$ L is an ideal of R , $I \subset L \subset P_s$, and $L_{P_r} \subset J_0\}$. $I \in \mathcal{I}$ so $\mathcal{I} \neq \emptyset$. By Zorn's Lemma $\mathcal{I}$ has a maximal element P . We will show that P is is a prime ideal of R . Let $x, y \in R - P$. $P + Rx \notin \mathcal{J}$, so $x \notin P_s$ or $R_{p_r}x \not\subset J_0$. If $x \notin P_s$, then since $J_0 \in \mathcal{J}$ we must have $R_{p_r}x \not\subset J_0$. Thus in all cases $R_{p_r}x \not\subset J_0$. Similarly $R_{p_r}y \not\subset J_0$. J_0 is a prime ideal of R_{p_r} implies $R_{p_r}xy \not\subset J_0$. Then $xy \in R - P$. This verifies that P is a prime ideal of R . Clearly $I \subset P \subset P_r \cap P_s$. <u>q.e.d.</u>

Let T be a torsion Abelian group. Then it is well known that $T = \bigoplus_{p \text{ prime}} T((p))$ where <u>$T((p))$</u> $= \{x \in T: p^n x = 0 \text{ for some } n \in N\}$. This says that all torsion Z-modules decompose into p-adic parts. How could this be generalized to domains? Using pZ to denote the prime ideal of Z generated by the prime integer p , then $T((p)) \cong T \otimes_Z Z_{pZ}$ and the latter by definition is just T_{pZ} , the localization of T at the prime ideal pZ . Thus the above statement could be given as $T \cong \bigoplus_{M \in \operatorname{mspec} Z} T_M$. We shall see that the domains with this decomposition property for torsion modules are exactly the h-local domains.

<u>Definition</u>: Let R be a domain. Then R is <u>h-local</u> if every non-zero element of R is an element of only finitely many maximal ideals of R and every non-zero prime ideal of R is a subset of only one maximal ideal of R .

Z is an h-local domain and every local domain is h-local. If k is a field and X and Y are indeterminants over k , then $R = k[X,Y]$ is not h-local. For the non-zero prime ideal RX is a subset of two distinct maximal ideals $RX + RY$ and $RX + R(Y + 1)$.

E. Matlis [22] defined h-local domains in 1964, the "h" designating "homological". In this original paper he showed that a domain R with quotient field Q is h-local if and only if $Q/R \cong \bigoplus_{M \in \operatorname{mspec} R} (Q/R)_M$ if and only if $H = \prod_{M \in \operatorname{mspec} R} H(M)$ where H is the completion of R in the R-topology and $H(M)$ is the completion of R_M in the R_M-topology. In this same paper he showed that R is h-local implies $T \cong \bigoplus_{M \in \text{mspec}R} T_M$ for all torsion R-modules T ; and in another paper two years later [23] he showed the converse. E. Matlis' proofs use homological algebra. Hence the proof given here for 2.6 is quite different from the original.

If T is an R-module and $M \in \text{mspec}R$, then define $\underline{T(M)}$ = $\{x \in T: \text{mspec}(\text{Ann}_R(x)) \subset \{M\}\}$. In particular if T is a Z-module and p is a prime integer, i.e., $Zp \in \text{mspec}Z$, then $T(Zp)$ is just the p-adic subgroup of the torsion subgroup of T . Thus this T(M) notation is consistent with the earlier notation $T((p))$ if one uses (p) to denote Zp .

<u>Theorem 2.6</u>: (E. Matlis [22], [23]). Let R be a domain. The following four statements are equivalent:

1. R is h-local.
2. $T = \bigoplus_{M \in \text{mspec}R} T(M)$ for all torsion R-modules T .
3. $T \cong \bigoplus_{M \in \text{mspec}R} T_M$ for all torsion R-modules T .
4. $T \cong \bigoplus_{M \in \text{mspec}R} T_M$ for all cyclic torsion R-modules T .

<u>Proof</u>: $1 \to 2$: Let T be a torsion R-module. Let $x \in T - \{0\}$. R is h-local implies $\text{mspec}(\text{Ann}_R(x))$ is finite. $Rx = R/\text{Ann}_R(x)$. By 2.4 Rx is a direct sum of non-zero indecomposable R-modules, each isomorphic to R/I_i for some proper ideal I_i of R . R is h-local and 2.5 implies $|\text{mspec}(I_i)| = 1$. Thus $Rx = \bigoplus_{i=1}^{n} Rx_i$ where $|\text{mspec}(\text{Ann}_R(x_i))| = 1$ for all $i = 1,\ldots,n$. This implies $x \in \sum_{M \in \text{mspec}R} T(M)$. If $y \in T(M_0) \cap \sum_{i=1}^{n} T(M_i)$ for distinct $M_0, M_1, \ldots, M_n \in \text{mspec}R$, then $\text{mspec}(\text{Ann}_R(y)) \subset \{M_0\} \cap \{M_1,\ldots,M_n\} = \emptyset$, and so $y = 0$. This shows that the desired sum is a direct sum.

$2 \to 3$: Let $M \in \text{mspec}R$. It suffices to show that $T(M) \cong T_M$ for T a torsion R-module. We wish to show that T(M) is an R_M-module. Let $x \in T(M) - \{0\}$

and $s \in R - M$. $\mathrm{Ann}_R(x) + Rs = R$, so $a + rs = 1$ for some $a \in \mathrm{Ann}_R(x)$ and $r \in R$. Thus $x = 1 \cdot x = (a + rs)x = s(rx)$. Define $\frac{1}{s}x = rx$. One checks that this makes $T(M)$ an R_M-module and $T(M)_M \cong T(M)$. Let $M' \in \mathrm{mspec}R - \{M\}$ and $x \in T(M) - \{0\}$. $\mathrm{Ann}_R(x) \not\subset M'$ so there exists $a \in \mathrm{Ann}_R(x) - M'$. In $T(M)_{M'}$, $x \otimes 1 = x \otimes \frac{a}{a} = xa \otimes \frac{1}{a} = 0$, so $T(M)_{M'} \cong \{0\}$. Using statement 2 and 2.3(2) $T_M = (\bigoplus_{P \in \mathrm{mspec}R} T(P))_M \cong \bigoplus_{P \in \mathrm{mspec}R} T(P)_M \cong T(M)_M \cong T(M)$.

3 → 4: Trivial.

4 → 1: Let I be a non-zero ideal of R . By statement 4,

$R/I \cong \bigoplus_{M \in \mathrm{mspec}R} (R/I)_M \cong \bigoplus_{M \in \mathrm{mspec}R} R_M/I_M$. $R_M/I_M \ncong \{0\}$ if and only if

$M \in \mathrm{mspec}\ (I)$. A cyclic module cannot decompose into a direct sum of infinitely many non-zero submodules, so $|\mathrm{mspec}\ (I)| < \infty$. Let P be a non-zero prime idea of R . Again by statement 4 , $R/P = \bigoplus_{M \in \mathrm{mspec}R} R_M/P_M$, and $R_M/P_M \ncong \{0\}$ if

and only if $M \in \mathrm{mspec}\ (P)$. R/P is indecomposable, so $|\mathrm{mspec}\ (P)| = 1$. By definition, R is then h-local. q.e.d.

The proof given here for 2.6 has the advantage that it gives an easy generalization to rings which are not necessarily domains. To do this one needs to generalize some definitions. We say a ring (possibly with zero divisors) is h-local if every regular element (non-zero divisor and non-zero) of R is an element of only finitely many maximal ideals of R and every prime ideal of R which is not a minimal prime is a subset of only one maximal ideal of R . We say that an R-module T is torsion if for all $x \in T$ there exists a regular element $r \in R$ such that $rx = 0$. Then 2.6 holds without the assumption that R is a domain.

Corollary 2.7: Let R be an h-local domain. Then:

1. If T is a torsion R-module and $M \in \mathrm{mspec}R$, then $T(M) \cong T_M$ and $T(M)$ is an R_M-module.
2. If T is a torsion R-module, $M, M' \in \mathrm{mspec}R$, and $M \neq M'$, then

$(T_M)_{M'} \cong \{0\}$.

3. If T is a torsion R_M-module where $M \in \operatorname{mspec} R$, then the set of R_M-submodules of T equals the set of R-submodules of T .

Proof:

1. This was shown in the proof of 2.6.
2. Again refering to the proof of 2.6 $(T_M)_{M'} \cong T(M)_{M'} \cong \{0\}$.
3. Clearly every R_M-submodule of T is an R-submodule of T . Suppose A is an R-submodule of T . Let $M' \in \operatorname{mspec} R - \{M\}$. By statement 2 , $T_{M'} \cong (T_M)_{M'} \cong \{0\}$, and so $A_{M'} \cong \{0\}$. By 2.6, $A = A(M) \oplus (\bigoplus_{M' \in \operatorname{mspec} R - \{M\}} A(M'))$ and $A(M') \cong A_{M'} \cong \{0\}$. Hence $A = A(M)$ and so A is an R_M-module by part 1 . q.e.d.

Definition: R is a locally almost maximal ring if R_M is an almost maximal ring for all $M \in \operatorname{mspec} R$.

Lemma 2.8: If R is a locally almost maximal ring and I is a proper ideal of R , then R/I is a locally almost maximal ring.

Proof: Let $M' \in \operatorname{mspec}(R/I)$. There exists $M \in \operatorname{mspec} R$ such that $M \supset I$ and $M' = M/I$. $(R/I)_{M'} = (R/I)_{M/I} \cong R_M/I_M$ and the latter is an almost maximal ring since R is a locally almost maximal ring. q.e.d.

Theorem 2.9: (W. Brandal [2]) . Let R be a domain. R is an almost maximal ring if and only if R is h-local and R is a locally almost maximal ring.

Proof: Suppose R is an almost maximal ring. Let $r \in R - \{0\}$. We wish to show that r is an element of only finitely many maximal ideals of R , so we may also assume that r is not a unit of R . R/Rr is a linearly compact R-module, hence a linearly compact R/Rr-module. Thus R/Rr is a maximal ring. By 1.6 R/Rr has only a finite number of maximal ideals, and so r is an element of only finitely many maximal ideals of R . Let P be a non-zero prime ideal of R . Then R/P is a maximal domain, so by 1.7 R/P is local. Thus P is a subset of only one maximal ideal of R . This shows that R is h-local.

Let $M \in \text{mspec}R$ and let J be a non-zero ideal of R_M. Define $I = J \cap R$ (viewing R and R_M as subsets of the quotient field of R). By 2.6 $R/I \cong (R/I)_M \oplus (\bigoplus_{M' \in \text{mspec}R - \{M\}} (R/I)_{M'})$. $(R/I)_M \cong R_M/I_M \cong R_M/R_M I \cong R_M/R_M(J \cap R) \cong R_M/J$. Since R is almost maximal, R/I is a linearly compact R-module. By 1.2(1) R_M/J is a linearly compact R-module, and hence R_M/J is a linearly compact R_M-module. This verifies that R is a locally almost maximal ring.

Conversely, suppose R is h-local and locally almost maximal. Let I be a non-zero ideal of R. We need to show that R/I is a linearly compact R-module. By 2.6 and 2.7(1), $R/I = \bigoplus_{M \in \text{mspec}R} R/I(M)$ and $R/I(M) \cong (R/I)_M \cong R_M/I_M$. By assumption R_M/I_M is a linearly compact R_M-module. By 2.7(3) R_M/I_M is a linearly compact R-module. Since R is h-local, I is a subset of only finitely many maximal ideals of R. Thus R/I is a finite direct sum of linearly compact R-modules, and so is linearly compact by 1.2(4). q.e.d.

Although we shall not need the following in our latter development, we discuss a few localization results. The property of a ring being maximal or almost maximal is not preserved by localizations. For consider k a field, X_1, X_2, X_3 indeterminants over k, $R = k[[X_1,X_2,X_3]]$, and $P = RX_2 + RX_3$. It can be shown that R is a maximal ring and $P \in \text{spec}R$, yet R_P is not an almost maximal ring [2]. The property of a ring being a maximal ring is preserved by localizing at a prime ideal if R is a valuation domain, as we shall see in 10.10(2), and the reader is refered to [7] for generalizations to valuation rings. If one restricts to localizations at maximal ideals, then one gets some affirmative results. By 1.6, if R is a maximal ring, then trivially R_M is a maximal ring for all $M \in \text{mspec}R$. By 2.9, if R is an almost maximal domain, then R is locally almost maximal. If R is a Noetherian ring and R is not a domain, then R is almost maximal if and only if R is maximal [2]. Thus it follows that if R is a Noetherian almost maximal ring, then R is locally a most maximal.

Section 3 Valuation Rings and Bezout Rings

Definition: R is a valuation ring if given any two elements of R then one divides the other.

Note that the definition does not require R to be a domain. In the literature this is sometimes refered to as a generalized valuation ring, and then a valuation ring is assumed to be a domain.

Lemma 3.1: The following three statements are equivalent:

1. R is a valuation ring.
2. The set of ideals of R is totally ordered with respect to set inclusion.
3. If $x_1,\ldots,x_n \in R$, then $\sum_{i=1}^{n} Rx_i = Rx_j$ for some $j \in \{1,\ldots,n\}$.

If R is a domain with quotient field Q, then the above statements are also equivalent to each of the following four statements:

4. Given any two elements of Q, then one R-divides the other.
5. The set of R-submodules of Q is totally ordered with respect to set inclusion.
6. If $x_1,\ldots,x_n \in Q$, then $\sum_{i=1}^{n} Rx_i = Rx_j$ for some $j \in \{1,\ldots,n\}$.
7. If $x \in Q - \{0\}$, then $x \in R$ or $1/x \in R$.

Proof: Straight forward.

It follows from 3.1 that valuation rings are local. Examples of valuation rings include fields, Z_{pZ}, and Z/p^nZ where p is a prime integer and $n \in N$. By a discrete rank one valuation domain is meant a valuation domain which is Noetherian and not a field. If R is a discrete rank one valuation domain and M is its maximal ideal, then it is easy to see that M is a principal ideal of R and the only non-trivial proper ideals of R are M^n for $n \in N$. Z_{pZ} is a discrete rank one valuation domain for p a prime integer. Valuation domains will be discussed in greater detail in section 10.

Lemma 3.2:

1. If R is a valuation ring and I is a proper ideal of R, then R/I is a valuation ring.
2. If R is a valuation ring and S is a multiplicatively closed subset of R, then R_S is a valuation ring.

Proof: Trivial.

The following result was done by D.T. Gill [7] in 1971 and independently by W. Brandal [2] in 1973.

Proposition 3.3: (D.T. Gill [7]) Let R be a valuation ring which is not a domain. Then R is an almost maximal ring if and only if R is a maximal ring.

Proof: Clearly if R is a maximal ring, then R is an almost maximal ring.

For the converse we begin by showing that if I is an ideal of R and R/I is a linearly compact R-module, then R/I^2 is a linearly compact R-module. Suppose $\{r_\alpha + I_\alpha\}_{\alpha \in X}$ is a family of cosets of submodules of R with the f.i.p. such that $I_\alpha \supset I^2$ for all $\alpha \in X$. We wish to show that this family has a non-empty intersection. If $I_\alpha \supset I$ for all $\alpha \in X$, then one gets the desired conclusion from the hypothesis that R/I is a linearly compact R-module. Thus assume there exists $\beta \in X$ such that $I \supsetneq I_\beta$. Let $y \in I - I_\beta$. Define $f: R/I \to R/I^2$ by $f(r + I) = ry + I^2$. f is an R-homomorphism and $f(R/I) = Ry/I^2$. Thus Ry/I^2 is a linearly compact R-module by 1.2(2). Let $Y = \{\alpha \in X: I_\alpha \subset I_\beta\}$ and let a prime denote modulo I^2. $\{r'_\alpha - r'_\beta + I'_\alpha\}_{\alpha \in Y}$ is a family of cosets of submodules of $Ry/I^2 = Ry'$ with the f.i.p.. Ry' is a linearly compact R-module implies there exists $x' \in \bigcap_{\alpha \in Y} r'_\alpha - r'_\beta + I'_\alpha$. It follows that $x + r_\beta \in \bigcap_{\alpha \in Y} r_\alpha + I_\alpha = \bigcap_{\alpha \in X} r_\alpha + I_\alpha$. This verifies that R/I^2 is a linearly compact R-module if R/I is a linearly compact R-module.

Suppose R is an almost maximal ring. Let $P = \bigcap \operatorname{spec} R$. Then P is the minimal prime ideal of R. $P \neq \{0\}$ since R is not a domain. P is the set of nilpotent elements of R, and so there exists $n \in P - \{0\}$ such that

$n^2 = 0$. R is an almost maximal implies R/Rn is a linearly compact R-module. By the last paragraph, $R \cong R/(Rn)^2$ is then a linearly compact R-module, and so R is a maximal ring. q.e.d.

Suppose R is a discrete rank one valuation domain and I is a non-zero ideal of R . Then R/I satisfies the descending chain condition on submodules. By 1.2(5) R/I is a linearly compact R- module. This verifies that a discrete rank one valuation domain is an almost maximal ring. In other words, a discrete rank one valuation domain is a P.I.D., and we know that a P.I.D. is an almost maximal ring.

The following gives a uniqueness of decomposition of modules into a direct sum of cyclics over valuation rings. Alternate proofs of this fact use exterior algebras.

Proposition 3.4: Let R be a valuation ring and A an R-module with $A = A_1 \oplus \ldots \oplus A_m = B_1 \oplus \ldots \oplus B_n$ where A_i and B_j are non-zero cyclic R-modules for all $i = 1, \ldots, m$ and $j = 1, \ldots, n$. Then $m = n$ and by a possible relabelling of the subscripts $A_i \cong B_i$ for all $i = 1, \ldots, m$.

Proof: Let M be the maximal ideal of R . The number of non-zero cyclic summands in a direct decomposition of A equals the dimension of A/MA as an R/M-vector space. This dimension is independent of the decomposition, and so $m = n$.

Let $A_i = Ra_i$ and $B_i = Rb_i$ for $a_i, b_i \in A$, $i = 1, \ldots, m$. By a possible relabelling of the subscripts we may assume that $\mathrm{Ann}_R(A_i) \subset \mathrm{Ann}_R(A_{i+1})$ and $\mathrm{Ann}_R(B_i) \subset \mathrm{Ann}_R(B_{i+1})$ for all $i = 1, \ldots, m-1$. For each $i = 1, \ldots, m$ $a_i = \sum_{j=1}^{m} r_{ij}b_j$ for some $r_{ij} \in R$. For elements of A let a bar denote modulo MA , and for elements of R let a bar denote modulo M . Then $\{\bar{a}_1,\ldots,\bar{a}_m\}$ and $\{\bar{b}_1,\ldots,\bar{b}_m\}$ are bases of A/MA as an R/M-vector space. Relative to these bases, the $m \times m$ matrix $(\bar{r}_{ij})^{\text{transpose}}$ is the matrix representations of the identity automorphism of A/MA. Thus $\det(\bar{r}_{ij}) \neq \bar{0}$, and so $\det(r_{ij})$ is a unit of R.

We wish to prove that $\mathrm{Ann}_R(A_i) \subset \mathrm{Ann}_R(B_i)$ for all $i = 1,\ldots,m$. Let $i_0 \in \{1,\ldots,m\}$. Since R is local and $\det(r_{ij})$ is a unit of R, there exists $i' \in \{i_0, i_0+1, \ldots, m\}$ and there exists $j' \in \{1,2,\ldots,i_0\}$ such that $r_{i'j'}$ is a unit of R. Then $\mathrm{Ann}_R(A_{i_0}) \subset \mathrm{Ann}_R(A_{i'}) = \mathrm{Ann}_R(a_{i'}) = \mathrm{Ann}_R(\sum_{j=1}^{m} r_{i'j}b_j) = \bigcap_{j=1}^{m} \mathrm{Ann}_R(r_{i'j}b_j) \subset \mathrm{Ann}_R(r_{i'j'}b_{j'}) = \mathrm{Ann}_R(b_{j'}) = \mathrm{Ann}_R(B_{j'}) \subset \mathrm{Ann}_R(B_{i_0})$. This verifies that $\mathrm{Ann}_R(A_i) \subset \mathrm{Ann}_R(B_i)$ for all $i = 1,\ldots,m$. Similarly $\mathrm{Ann}_R(B_i) \subset \mathrm{Ann}_R(A_i)$ for all $i = 1,\ldots,m$. Thus $\mathrm{Ann}_R(A_i) = \mathrm{Ann}_R(B_i)$ and so $A_i \cong B_i$ for all $i = 1,\ldots,m$. q.e.d.

Definition: R is a Bezout ring if every finitely generated ideal of R is cyclic.

Examples of Bezout rings include valuation rings and P.I.D.'s.

Lemma 3.5:

1. If R is a Bezout ring and I is a proper ideal of R, then R/I is a Bezout ring.
2. If R is a Bezout ring and S is a multiplicatively closed subset of R, then R_S is a Bezout ring.

Proof: Straight forward.

Lemma 3.6: If R is a local Bezout ring, then R is a valuation ring.

Proof: Assume R is a local Bezout ring and $a,b \in R$, not both elements zero. Since R is a Bezout ring, there exists $g \in R$ and that $Rg = Ra + Rb$. Then $g = r_1a + r_2b$, $a = s_1g$, and $b = s_2g$ for some $r_1, r_2, s_1, s_2 \in R$. $g = (r_1s_1 + r_2s_2)g$, $r_1s_1 + r_2s_2$ is a unit of R, and so $Rg = Ra$ or Rb. Thus R is a valuation ring by 3.1. q.e.d.

Definition: R is a Prufer ring if R_M is a valuation ring for all $M \in \mathrm{mspec}\,R$.

Prufer rings are also called arithmetic rings in the literature. Examples of Prufer rings include Dedekind domains and valuation rings. For domains, this definition of Prufer rings is equivalent to all the commonly used definitions. However, for rings which are not domains, there are non-equivalent definitions for Prufer rings used in the literature. A good reference for Prufer domains is the text by R. Gilmer [9] .

Corollary 3.7: If R is a Bezout ring, then R is a Prufer ring.

Proof: This follows directly from 3.5(2) and 3.6.

The converse of 3.7 is not true. From number theory there exist Dedekind domains which are not P.I.D.'s . It follows from the definitions that such domains are Prufer domains but not Bezout domains. The following is a partial converse.

Proposition 3.8: (Y. Hinohara [11]) If R is a semilocal Prufer ring, then R is a Bezout ring.

Proof: Suppose R is a Prufer ring with $\text{mspec}R = \{M_1,\ldots,M_n\}$. We assume the M_i's are distinct and $n \geq 2$. Let I be a finitely generated ideal of R , say $I = \sum_{k=1}^{m} Rx_k$ for $x_k \in I$. For each $i = 1, \ldots, n$ there exists $r_i \in (M_1 \cap \ldots \cap M_{i-1} \cap M_{i+1} \cap \ldots \cap M_n) - M_i$. For each $i = 1,\ldots,n$ $I_{M_i} = \sum_{k=1}^{m} R_{M_i} x_k$ by 2.3(2) , and since R_{M_i} is a valuation ring there exists $k_i \in \{1,\ldots,m\}$ such that $I_{M_i} = R_{M_i} x_{k_i}$ by 3.1 . Define $x = \sum_{j=1}^{n} r_j x_{k_j}$. For each $i = 1,\ldots,n$

$$R_{M_i} x = R_{M_i} r_i x_{k_i} \left(1 + \sum_{\substack{j=1 \\ j\neq i}}^{m} \frac{r_j}{r_i} \frac{x_{k_j}}{x_{k_i}}\right) = R_{M_i} r_i x_{k_i} = R_{M_i} x_{k_i} = I_{M_i} \text{ since}$$

$\sum_{\substack{j=1 \\ j\neq i}}^{m} \frac{r_j}{r_i} \frac{x_{k_j}}{x_{k_i}} \in M_i R_{M_i}$. By 2.1 $Rx = \bigcap_{i=1}^{n} (R_{M_i} x)^c = \bigcap_{i=1}^{n} (I_{M_i})^c = I$, and so I is cyclic. q.e.d.

Lemma 3.9: Let R be a domain. R is a Bezout domain if and only if every finitely generated torsion-free R-module is a free R-module.

Proof: The one implication is trivial. Conversely, suppose R is a Bezout domain with quotient field Q. If A is an R-module, then we define $\operatorname{rank} A$ to be the dimension of $A \otimes_R Q$ as a Q-vector space. Clearly the rank of a torsion R-module is zero. Suppose F is a finitely generated torsion-free R-module. We will show that F is a free R-module by induction on $\operatorname{rank} F$. If $\operatorname{rank} F$ is 0 or 1, then clearly F is a free R-module. Suppose $\operatorname{rank} F = n \geq 2$ and the result is known for smaller ranks. Let $x \in F - \{0\}$. Then $n - 1 = \operatorname{rank}(F/Rx) = \operatorname{rank}((F/Rx)/t(F/Rx))$ where $t(F/Rx)$ is the torsion submodule of F/Rx. By the induction hypothesis, $(F/Rx)/t(F/Rx)$ is a free R-module. The composition $F \to F/Rx \to (F/Rx)/t(F/Rx)$ is an epimorphism, and so there exists an R-module F' such that $F \cong F' \oplus ((F/RX)/t(F/Rx))$. F' is a finitely generated torsion-free R-module and $\operatorname{rank} F' = 1 < n$. Thus F' is a free R-module, and so F is a free R-module. q.e.d.

Although we shall not use it, 3.9 is comparable to the well-known result that for a domain R, R is Prufer if and only if every finitely generated torsion-free R-module is a projective R-module.

Section 4 Basic Facts About FGC Rings and the Local Case

Recall that R is an FGC ring if every finitely generated R-module decomposes into a direct sum of cyclic submodules. An alternate way of looking at FGC rings is in terms of bases. If A is an R-module and $\{b_1,\ldots,b_n\} \subset A - \{0\}$, then $\{b_1,\ldots,b_n\}$ is a basis of A if $A = Rb_1 \oplus \ldots \oplus Rb_n$. Then R is an FGC ring if and only if every finitely generated R-module has a basis.

Lemma 4.1:

1. Suppose $R = \prod_{i=1}^{n} R_i$. Then R is an FGC ring if and only if R_i is an FGC ring for all $i = 1,\ldots,n$.
2. If R is an FGC ring and I is a proper ideal of R, then R/I is an FGC ring.
3. If R is an FGC ring and S is a multiplicatively closed subset of R, then R_S is an FGC ring.

Proof: Straight forward.

One form of the Fundamental Theorem of Abelian Groups says that P.I.D.'s are FGC rings. The first theorem about FGC rings beyond this was in 1952 by I. Kaplansky [15]. It stated that almost maximal valuation domains are FGC domains. The proof given here is taken from this original source with a minor error corrected as in the paper by R.B. Warfield [32]. As is apparent, the assumption that R is a domain is not necessary. The next few facts then give a characterization of the local FGC rings.

Definition: Let B be a submodule of the R-module A. Then B is a pure submodule of A if $rB = rA \cap B$ for all $r \in R$.

For R-modules $A \supset B \supset C$, if C is pure in A, then C is pure in B.

Theorem 4.2:(I. Kaplansky [15]) If R is an almost maximal valuation ring, then R is an FGC ring.

Proof: Let A be a finitely generated R-module. We use induction on n, the minimal number of generators of A. For $n = 1$, the required decomposition is trivial. Assume A is generated by n elements and not by fewer than n elements, and assume every R-module with fewer than n generators decompose into a direct sum of cyclics.

Let M be the maximal ideal of R and let $J = \mathrm{Ann}_R(A)$. A/MA is a finite dimensional R/M-vector space, say with basis $\{b_1,\dots,b_m\}$. Choose $x_i \in A$ such that $x_i + MA = b_i$ for $i = 1,\dots,m$. Then $A = \sum_{i=1}^{m} Rx_i$ and $m = n$. Suppose for all $i = 1,\dots,n$ one could have chosen x_i such that $\mathrm{Ann}_R(x_i) \supsetneq J$. Since R is a valuation ring $J \supset \bigcap_{i=1}^{n} \mathrm{Ann}_R(x_i) \supsetneq J$. This cannot happen, so for some i , say $i = 1$, we must have the condition that if $x \in A$ and $x + MA = b_1$, then $\mathrm{Ann}_R(x) = J$. In particular $\mathrm{Ann}_R(x_1) = J$.

We claim that Rx_1 is a pure submodule of A . Let $r \in R$ and we wish to show that $rRx_1 \supset rA \cap Rx_1$. Let $a \in A$ and suppose $ra = sx_1$ for $s \in R$. If $r|s$ with $s = rt$ for $t \in R$, then $ra = sx_1 = rtx_1 \in rRx_1$, showing the desired purity. On the other hand suppose $r \nmid s$. R is a valuation ring implies $s|r$ with $r = sp$ for some $p \in M$. Let $x' = x_1 - pa$. Then $x' + MA = b_1$ and $rx' = r(x_1 - pa) = rx_1 - rpa = rx_1 - psx_1 = rx_1 - rx_1 = 0$. By the last paragraph $r \in \mathrm{Ann}_R(x') = J$, $ra = 0 \in rRx_1$, which shows that Rx_1 is a pure submodule of A .

A/Rx_1 is generated by $n - 1$ elements, so by the inductive hypothesis $A/Rx_1 = \bigoplus_{j=1}^{k} Rz_j$. Let $j \in \{1,\dots,k\}$. We wish to find a $y_j \in A$ such that $y_j + Rx_1 = z_j$ and $\mathrm{Ann}_R(y_j) = \mathrm{Ann}_R(z_j)$. Let $I = \mathrm{Ann}_R(z_j)$ and choose $y \in A$ such that $y + Rx_1 = z_j$. For $\alpha \in I$, $\alpha y \in Rx_1$. Rx_1 is a pure submodule of A , so $\alpha y = \alpha x_\alpha$ for some $x_\alpha \in Rx_1$. For $\alpha \in I$ let $I_\alpha = \{r \in R\colon r\alpha \in J\}$. Then $I_\alpha x_1 = \{x \in Rx_1\colon \alpha x = 0\}$, for $r \in I_\alpha$ if and only if $r\alpha \in J = \mathrm{Ann}_R(x_1)$ if and only if $r\alpha x_1 = 0$ if and only if $rx_1 \in \{x \in Rx_1\colon \alpha x = 0\}$.

Consider $\{x_\alpha + I_\alpha x_1\}_{\alpha \in I}$. We claim this is a family of cosets of sub-

modules of Rx_1 with the f.i.p. . Let $\alpha,\beta \in I$. Without loss of generality we may assume $\alpha|\beta$, i.e., $\beta = \alpha t$ for $t \in R$. $\beta(x_\alpha - x_\beta) = \beta x_\alpha - \beta x_\beta = t\alpha x_\alpha - \beta x_\beta = t\alpha y - \beta y = \beta y - \beta y = 0$. Thus $x_\alpha - x_\beta \in \{x \in Rx_1 : \beta x = 0\} = I_\beta x_1$, and so $x_\alpha + I_\alpha x_1 = (x_\alpha + I_\alpha x_1) \cap (x_\beta + I_\beta x_1)$. It follows that $\{x_\alpha + I_\alpha x_1\}_{\alpha\in I}$ has the f.i.p. . $Rx_1 \cong R/J$. If R is a domain or $J = \{0\}$, then a short calculation shows that Rx_1 is a direct summand of $Rx_1 + Ry$, and so the desired y_j exists. Thus we suppose R is not a domain or $J \neq \{0\}$. Since R is almost maximal, and possibly using 3.3 one sees that Rx_1 is linearly compact R-module. Thus there exists $x_0 \in \bigcap_{\alpha\in I} x_\alpha + I_\alpha x_1$. Define $y_j = y - x_0$. We claim this is the required y_j .
$y_j + Rx_1 = y - x_0 + Rx_1 = y + Rx_1 = z_j$, since $x_0 \in Rx_1$. $\mathrm{Ann}_R(y_j) \subset \mathrm{Ann}_R(z_j) = I$ On the other hand suppose $\alpha \in I$. $x_0 \in x_\alpha + I_\alpha x_1$ implies $x_0 = x_\alpha + r_\alpha x_1$ for some $r_\alpha \in I_\alpha$. Then $\alpha y_j = \alpha(y - x_0) = \alpha(y - x_\alpha - r_\alpha x_1) = \alpha y - \alpha x_\alpha - \alpha r_\alpha x_1 = -\alpha r_\alpha x_1$. $r_\alpha \in I_\alpha$ implies $r_\alpha \alpha \in J = \mathrm{Ann}_R(x_1)$, so $-\alpha r_\alpha x_1 = 0$. Hence $\alpha y_j = 0$, $\alpha \in \mathrm{Ann}_R(y_j)$, and so $\mathrm{Ann}_R(y_j) = \mathrm{Ann}_R(z_j)$. This verifies that the required y_j exist.

We claim that $A = Rx_1 \oplus (\bigoplus_{j=1}^{k} Ry_j)$. Clearly $A = Rx_1 + \sum_{j=1}^{k} Ry_j$. By modding Rx_1 and using $\mathrm{Ann}_R(y_j) = \mathrm{Ann}_R(z_j)$ it follows that A is the direct sum of Rx_1 and $\sum_{j=1}^{k} Ry_j$. By modding Rx_1 it follows that $\sum_{j=1}^{k} Ry_j$ is a direct sum. This completes the induction augument. q.e.d.

The converse appears in the literature in at least three places. First the domain case was done by E. Matlis [23] in 1966, in which case the following 4.3 is trivial. Then in 1971 D. Gill [7] and J.P. Lafon [18] independently proved the given general case. The arguments of E. Matlis and D. Gill used homological algebra, and we present the more elementary proof of J.P. Lafon.

Proposition 4.3: If R is a local FGC ring, then R is a valuation ring.

Proof: Suppose R is a local FGC ring with maximal ideal M. Assume R is not a valuation ring, and so there exists $x_0, y_0 \in R$ with neither element dividing the other. If $Rx_0 + Ry_0$ is a cyclic ideal, then by taking linear combinations and considering coefficients, one deduces that x_0 or y_0 divides the other. Thus assume $Rx_0 + Ry_0$ is not cyclic. By considering R/M-dimensions, $Rx_0 + Ry_0$ cannot be a direct sum of more than two non-zero submodules. Since R is an FGC ring, $Rx_0 + Ry_0 = Rx \oplus Ry$ for some $x, y \in R - \{0\}$. If $x \notin M$ or $y \notin M$ then $Rx \oplus Ry$ is not a direct sum. Thus we assume $x, y \in M$. Define $A = R^2/R(x,y)$. Let $b_1 = (1,0) + R(x,y)$ and $b_2 = (0,1) + R(x,y)$. Then $A = Rb_1 + Rb_2$. Since R is an FGC ring and by considering R/M-dimensions we must have $A = Ra_1 \oplus Ra_2$ for some $a_1, a_2 \in A - \{0\}$. There exist $r_{ij}, s_{ij} \in R$ such that

$$\begin{pmatrix} r_{11} & r_{12} \\ r_{21} & r_{22} \end{pmatrix} \begin{pmatrix} a_1 \\ a_2 \end{pmatrix} = \begin{pmatrix} b_1 \\ b_2 \end{pmatrix} \quad \text{and} \quad \begin{pmatrix} s_{11} & s_{12} \\ s_{21} & s_{22} \end{pmatrix} \begin{pmatrix} b_1 \\ b_2 \end{pmatrix} = \begin{pmatrix} a_1 \\ a_2 \end{pmatrix} .$$

For elements of R let a bar denote modulo M, and for elements of A let a bar denote modulo MA. Then $\{\bar{b}_1, \bar{b}_2\}$ and $\{\bar{a}_1, \bar{a}_2\}$ are bases of $\bar{A}$ as an $\bar{R}$-vector space. $(\bar{r}_{ij})$ and $(\bar{s}_{ij})$ are inverse matrices of each other over the field $\bar{R}$, and in particular $\det(\bar{r}_{ij}) \neq 0$ and $\det(\bar{s}_{ij}) \neq 0$.

Let $\alpha_1 = r_{11}x + r_{21}y$ and $\alpha_2 = r_{12}x + r_{22}y$. From the definition of A, we have $xb_1 = -yb_2$, $x(r_{11}a_1 + r_{12}a_2) = -y(r_{21}a_1 + r_{22}a_2)$, $(r_{11}x + r_{21}y)a_1 = -(r_{12}x + r_{22}y)a_2$. By the direct sum decomposition, these last expressions are zero, showing $\alpha_i \in \mathrm{Ann}_R(a_i)$ for $i = 1,2$.

If $\alpha_1 = 0$, then $r_{11}x + r_{21}y = 0$. $\det(\bar{r}_{ij}) \neq 0$ implies r_{11} or r_{21} is a unit of R, and so either x or y divides the other. This contradicts $Rx \oplus Ry$ is a direct sum. Thus $\alpha_1 \neq 0$ and similarly $\alpha_2 \neq 0$.

$\alpha_1 a_1 = 0$ gives $\alpha_1(s_{11}b_1 + s_{12}b_2) = 0$, $(\alpha_1 s_{11}, \alpha_1 s_{12}) \in R(x,y)$, and so there exists $s \in R$ such that $\alpha_1 s_{11} = sx$ and $\alpha_1 s_{12} = sy$. $\det(\bar{s}_{ij}) \neq 0$ implies s_{11} or s_{12} is a unit of R. Without loss of generality assume s_{11} is a unit of R. Then $sy = \alpha_1 s_{12} = (s_{11}{}^{-1} sx)s_{12} \in Rx \cap Ry = \{0\}$. Therefore $s \in M$ and $s_{12} \in M$ since $\alpha_1 \neq 0$ and $y \neq 0$. The product $(\bar{s}_{ij})(\bar{r}_{ij})$ is the identity matrix implies $s_{11}r_{11} + s_{12}r_{21}$ is a unit of R, and so r_{11} is a unit of R. $\alpha_1 s_{11} = sx$ gives $(r_{11}x + r_{21}y)s_{11} = sx$, $(r_{11}s_{11} - s)x = -r_{21}s_{11}y \in Rx \cap Ry = \{0\}$. But $r_{11}s_{11} - s$ is a unit of R, so $x = 0$. This contradicts $x \neq 0$. <u>q.e.d.</u>

<u>Theorem 4.4</u>: (D. Gill [7], J.P. Lafon [18]) If R is a local FGC ring, then R is an almost maximal valuation ring.

<u>Proof:</u> Suppose R is a local FGC ring. By 4.3 R is a valuation ring. Let M be the maximal ideal of R. Suppose I is a non-zero ideal of R. To show R is an almost maximal ring we must show that R/I is a linearly compact R-module.

Suppose $\{x'_\alpha + I'_\alpha\}_{\alpha \in X}$ is a family of cosets of submodules of R/I with the f.i.p.. Let $x_\alpha \in R$ such that $x_\alpha + I = x'_\alpha$ and let I_α be an ideal of R such that $I_\alpha/I = I'_\alpha$. Then $\{x_\alpha + I_\alpha\}_{\alpha \in X}$ is a family of cosets of submodules of R with the f.i.p., $\bigcap_{\alpha \in X} I_\alpha \supset I \neq \{0\}$, and it suffices to show that

$\bigcap_{\alpha \in X} x_\alpha + I_\alpha \neq \emptyset$.

Let $J = \bigcap_{\alpha \in X} I_\alpha$. If there exists a $\beta \in X$ such that $J = I_\beta$, then

$\bigcap_{\alpha \in X} x_\alpha + I_\alpha = x_\beta + I_\beta \neq \emptyset$, completing the proof. Thus we assume $I_\alpha \supsetneqq J$ for all $\alpha \in X$.

Let $\alpha \in X$. Choose $y_\alpha \in I_\alpha - J$ and choose $\beta \in X$ such that $Ry_\alpha \supset I_\beta$. Then $x_\alpha + I_\alpha \supset x_\beta + Ry_\alpha \supset x_\beta + I_\beta$. Replacing $x_\alpha + I_\alpha$ by $x_\beta + Ry_\alpha$ amounts to assuming that I_α is a cyclic ideal of R. Thus we consider $\{x_\alpha + Ry_\alpha\}_{\alpha \in X}$

a family of cosets of submodules of R with the f.i.p., $\bigcap_{\alpha\in X} Ry_\alpha = J \neq \{0\}$, $Ry_\alpha \supsetneq J$ for all $\alpha \in X$, and it suffices to show that $\bigcap_{\alpha\in X} x_\alpha + Ry_\alpha \neq \emptyset$. If for some $\alpha \in X$, $y_\alpha \notin M$, then $x_\alpha + Ry_\alpha = R$ and so we may discard this coset. Thus we will also assume that $y_\alpha \in M$ for all $\alpha \in X$.

Choose $j \in J - \{0\}$. For all $\alpha \in X$, $Ry_\alpha \supsetneq J \supset Rj$, so there exists $q_\alpha \in M$ such that $q_\alpha y_\alpha = j$. Define B to be the submodule of R^2 given by $B = (Rj \oplus Rj) + \sum_{\alpha\in X} R(q_\alpha, -x_\alpha q_\alpha)$, and define $A = R^2/B$. Define $b_1,b_2 \in A$ by $b_1 = (1,0) + B$ and $b_2 = (0,1) + B$.

We claim that $\mathrm{Ann}_R(A) = Rj$. The one inclusion is obvious. Suppose $r \in \mathrm{Ann}_R(A)$. Then $0 = rb_2 = (0,r) + B$. Thus there exist $s_1,s_2 \in Rj$, $\{\alpha_1,\ldots,\alpha_n\} \subset X$, $r_i \in R$ such that $(0,r) = (s_1,s_2) + \sum_{i=1}^{n} r_i(q_{\alpha_i}, -x_{\alpha_i} q_{\alpha_i})$. By the f.i.p. , there exists $b \in \bigcap_{i=1}^{n} x_{\alpha_i} + Ry_{\alpha_i}$. For each i , $b = x_{\alpha_i} + t_i y_{\alpha_i}$ for some $t_i \in R$. Hence $(0,r) = (s_1,s_2) + \sum_{i=1}^{n} r_i(q_{\alpha_i}, -(b - t_i y_{\alpha_i})q_{\alpha_i}) =$ $(s_1,s_2') + \sum_{i=1}^{n} r_i(q_{\alpha_i}, -bq_{\alpha_i})$ where $s_2' = s_2 + \sum_{i=1}^{n} r_i t_i y_{\alpha_i} q_{\alpha_i} = s_2 + \sum_{i=1}^{n} r_i t_i j \in Rj$. By equating components, one sees that $r \in Rj$. This verifies the claim that $\mathrm{Ann}_R(A) = Rj$.

R is an FGC ring, so A is a direct sum of cyclic submodules. We have $B \subset MR^2$, so A/MA is a two dimensional R/M-vector space. Hence A must be a direct sum of exactly two non-zero cyclic submodules. Suppose $A = Ra_1 \oplus Ra_2$ for $a_1,a_2 \in A - \{0\}$. Thus $Rj = \mathrm{Ann}_R(A) = \mathrm{Ann}_R(a_1) \cap \mathrm{Ann}_R(a_2)$. Since R is a valuation ring, we may without loss of generality assume $\mathrm{Ann}_R(a_1) = Rj$. There exist $r_{ik}, s_{ik} \in R$ such that

$$\begin{pmatrix} r_{11} & r_{12} \\ r_{21} & r_{22} \end{pmatrix} \begin{pmatrix} a_1 \\ a_2 \end{pmatrix} = \begin{pmatrix} b_1 \\ b_2 \end{pmatrix} \quad \text{and} \quad \begin{pmatrix} s_{11} & s_{12} \\ s_{21} & s_{22} \end{pmatrix} \begin{pmatrix} b_1 \\ b_2 \end{pmatrix} = \begin{pmatrix} a_1 \\ a_2 \end{pmatrix} .$$

For elements of R let a bar denote modulo M, and for elements of A let a bar denote modulo MA. $\{\bar{a}_1, \bar{a}_2\}$ and $\{\bar{b}_1, \bar{b}_2\}$ are bases of $\bar{A}$ as an $\bar{R}$-vector space. $(\bar{r}_{ik})$ and $(\bar{s}_{ik})$ are inverse matrices of each other over the field $\bar{R}$, and in particular $\det(\bar{s}_{ik}) \neq 0$. Thus $\det(s_{ik})$ is a unit of R. By possibly changing the generators a_1 and a_2 by unit factors, we may assume $\det(s_{ik}) = 1$ Thus

$$\begin{pmatrix} b_1 \\ b_2 \end{pmatrix} = \begin{pmatrix} s_{11} & s_{12} \\ s_{21} & s_{22} \end{pmatrix}^{-1} \begin{pmatrix} a_1 \\ a_2 \end{pmatrix} = \begin{pmatrix} s_{22} & -s_{12} \\ -s_{21} & s_{11} \end{pmatrix} \begin{pmatrix} a_1 \\ a_2 \end{pmatrix} .$$

$\det(s_{ik}) = 1$ and R is local implies s_{21} or s_{22} is a unit of R.

Case 1: Suppose s_{21} is a unit of R. Let $x = -s_{21}^{-1} s_{22}$. We claim that $x \in \bigcap_{\alpha \in X} x_\alpha + Ry_\alpha$. Let $\alpha \in X$. By the definition of A, $q_\alpha b_1 - x_\alpha q_\alpha b_2 = 0$, $q_\alpha(s_{22}a_1 - s_{12}a_2) - x_\alpha q_\alpha(-s_{21}a_1 + s_{11}a_2) = 0$, $(q_\alpha s_{22} + x_\alpha q_\alpha s_{21})a_1 = 0$, $q_\alpha s_{22} + x_\alpha q_\alpha s_{21} \in \mathrm{Ann}_R(a_1) = Rj = Rq_\alpha y_\alpha$, and $q_\alpha(-s_{21}^{-1} s_{22}) - x_\alpha q_\alpha \in Rq_\alpha y_\alpha$. There exists $r \in R$ such that $q_\alpha x - x_\alpha q_\alpha = rq_\alpha y_\alpha$ and so $q_\alpha(x - x_\alpha - ry_\alpha) = 0$. $q_\alpha y_\alpha = j \neq 0$ implies $y_\alpha \notin \mathrm{Ann}_R(q_\alpha)$, and so $\mathrm{Ann}_R(q_\alpha) \subset Ry_\alpha$. Thus $x - x_\alpha - ry_\alpha \in Ry_\alpha$ and so $x \in x_\alpha + Ry_\alpha$, completing the proof in Case 1.

Case 2: Suppose s_{22} is a unit of R. Proceeding as in Case 1, $q_\alpha(s_{22} + x_\alpha s_{21} + ry_\alpha) = 0$ for some $r \in R$. $q_\alpha \neq 0$ implies $s_{22} + x_\alpha s_{21} + ry_\alpha \in M$. But s_{22} is a unit of R and $y_\alpha \in M$ implies $x_\alpha s_{21}$

is a unit of R, and so s_{21} is a unit of R. Apply Case 1 to complete the proof. q.e.d.

Theorem 4.5: (D. Gill [7], J.P. Lafon [18]) Let R be a local ring. Then R is an FGC ring if and only if R is an almost maximal valuation ring.

Proof: 4.2 and 4.4 .

Lemma 4.6: If R is an FGC ring, then R is a locally almost maximal Bezout ring.

Proof: We first show that R is a Bezout ring. Let I be a finitely generated ideal of R. R is an FGC ring implies I is finite direct sum of cyclic ideals. By induction it suffices to consider the case $I = Rx \oplus Ry$ for $x,y \in R$ Let $M \in \text{mspec}R$. By 4.1 R_M is an FGC ring, and so by 4.3 R_M is a valuation ring. $I_M = R_Mx \oplus R_My$ and R_M is a valuation ring implies $R_Mx = \{0\}$ or $R_My = \{0\}$. In either case $I_M = R_M(x + y)$. Using 2.1(1) $I = R(x + y)$, I is cyclic, and so R is a Bezout ring.

If $M \in \text{mspec}R$, then R_M is an FGC ring by 4.1, and so R_M is an almost maximal ring by 4.4 . Hence R is a locally almost maximal ring. q.e.d.

Section 5 Further Facts about FGC Rings and Torch Rings

In the last section the local FGC rings were characterized as the almost maximal valuation rings. As an easy consequence, FGC rings were seen to be locally almost maximal Bezout rings. This section begins with showing two other classes of rings are FGC rings, namely almost maximal Bezout domains and torch rings are FGC rings. That almost maximal Bezout domains are FGC rings was first seen by W. Brandal [2] in 1973, and independently by T. Shores and R. Wiegand [28] in 1974. The proofs were the same, being an easy consequence of the follow result, due to E. Matlis [23] .

Theorem 5.1: (E. Matlis [23]) If R is an h-local Prufer domain and R is a locally almost maximal ring, then every finitely generated torsion R-module decomposes into a direct sum of cyclic submodules.

Proof: Assume R satisfies the given hypotheses and T is a finitely generated torsion R-module. Using R is h-local, 2.6 and 2.7(1) $T = \bigoplus_{M \in \mathrm{mspec}R} T(M)$ and $T(M)$ is an R_M-module. Since T is finitely generated, there exists a finite set $\{M_1,\ldots,M_n\} \subset \mathrm{mspec}R$ such that $T = \bigoplus_{i=1}^{n} T(M_i)$. Let $i \in \{1,\ldots,n\}$. R is a Prufer ring and R is a locally almost maximal ring implies R_{M_i} is an almost maximal valuation ring. $T(M_i)$ is a finitely generated R_{M_i}- module, so by 4.2 $T(M_i)$ is an R_{M_i}-direct sum of cyclic submodules. By 2.7(3) the set of R_{M_i}-submodules of $T(M_i)$ equals the set of R-submodules of $T(M_i)$. Therefore $T(M_i)$ is an R-direct sum of cyclic submodules, and so T is the required direct sum of cyclic submodules. q.e.d.

Theorem 5.2: (W. Brandal [2], T. Shores and R. Wiegand [28]) If R is an almost maximal Bezout domain, then R is an FGC domain .

Proof: Suppose R is an almost maximal Bezout domain and A is a finitely generated R-module with $t(A)$ its torsion submodule. $A/t(A)$ is a finitely generated torsion-free R-module, and so by 3.9 $A/t(A)$ is a free R-module. Thus $A \cong A/t(A) \oplus t(A)$. By 2.9 R is an almost maximal ring implies R is h-local and locally almost maximal. By 3.7 R is a Bezout ring implies R is a Prufer ring. $t(A)$ is a finitely generated torsion R-module implies $t(A)$ is a direct sum of cyclic submodules by 5.1. Hence A is a direct sum of cyclic submodules, and so R is an FGC ring. q.e.d.

If R is a P.I.D., then clearly R is a Bezout domain. It has been remarked in section one that P.I.D.'s are almost maximal rings. Thus P.I.D.'s are almost maximal Bezout domains. The last theorem then includes as a special case the Fundamental Theorem of Abelian Groups, namely that P.I.D.'s are FGC rings.

An R-module is said to be a uniserial R-module if its set of submodules is totally ordered by set inclusion. In other words, a uniserial module is one in which any two submodules are comparable.

Definition: R is a torch ring if R satisfies all of the following:

1. R is not local,
2. R has a unique minimal prime ideal P and P is a non-zero uniserial R-module,
3. R/P is an h-local domain, and
4. R is a locally almost maximal Bezout ring.

Torch rings were first considered by T. Shores and R. Wiegand [28]. The name "torch ring" was suggested by P. Vamos [31], and of course the name refers to the description of the lattice of ideals (see 2. and 5. below). Condition 1. is included in the definition to exclude the possibility of a torch ring being an almost maximal valuation ring.

Lemma 5.3: Let R be a torch ring and let P be the minimal prime ideal of R. Then R satisfies all of the following conditions:

5. Every ideal of R contained in P is comparable to every ideal of R ,
6. $P^2 = \{0\}$,
7. there exists a unique $M \in \text{mspec}R$ such that $P \cong P_M$, and $P_{M'} \cong \{0\}$, for all $M' \in \text{mspec}R - \{M\}$, and
8. P is a torsion divisible R/P-module.

Proof:

5. If $r \in R - P$ and $p \in P$, then $R_M r \supset R_M p$ for all $M \in \text{mspec}R$ since R is a Bezout ring. By 2.1 $Rr = \bigcap_{M \in \text{mspec}R} (R_M r)^c \supset \bigcap_{M \in \text{mspec}R} (R_M p)^c = Rp$.
The required statement follows since P is a uniserial R-module.

6. Suppose $P^2 \neq \{0\}$. Then there exists $p \in P$ such that $Pp \neq \{0\}$. Thus $\text{Ann}_R(p) \not\supset P$. By (5), $\text{Ann}_R(p) \subset P$. Thus there exists an epimorphism $Rp \longrightarrow R/P$. $Rp \subset P$ are uniserial R-modules, so R/P is a uniserial R-module. But this contradicts $|\text{mspec}(R/P)| = |\text{mspec}R| > 1$.

7. $P^2 = \{0\}$ implies P is an R/P-module in the obvious manner. If $p \in P$ and p is not a torsion R/P-element of P , then $Rp = (R/P)p \cong R/P$ again contradicting Rp is uniserial and $|\text{mspec}(R/P)| > 1$. Thus P is a torsion R/P-module. By (3) R/P is an h-local domain, and so by 2.6 $P = \bigoplus_{M \in \text{mspec}(R/P)} P(M)$
P is a uniserial R-module by (2) and hence $P = P(M)$ for some unique $M \in \text{mspec}(R/P)$ and $P(M') = \{0\}$ for all $M' \in \text{mspec}(R/P) - \{M\}$. By the bijection between mspecR and mspec(R/P) one obtains the required statement.

8. In the proof of (7) we saw that P is a torsion R/P-module. Let $r \in R - P$ and $p \in P$. By (5) $Rr \supset Rp$ and so $p = sr$ for some $s \in R$. P is a prime ideal implies $s \in P$. Hence P is a divisible R/P-module. q.e.d.

Theorem 5.4: (T. Shores and R. Wiegand [28]) If R is a torch ring, then R is an FGC ring.

Proof: Suppose R is a torch ring. Let P be the unique minimal prime ideal of R and let M be the unique maximal ideal of R such that $P_M \cong P$.

Let A be a finitely generated R-module, and let $I = \mathrm{Ann}_R(A)$. If $I \supset P$, then A is a finitely generated R/P-module in the obvious manner. R/P is an almost maximal Bezout domain by the definition of torch ring, 3.5, 2.8, and 2.9 . By 5.2 A is a direct sum of cyclic R/P-modules, and so A is the desired direct sum of cyclic R-modules. Thus it suffices to consider the case where $I \not\supset P$. By 5.3(5) we may assume that $P \supsetneqq I$.

Since A is a finitely generated R-module and 5.3(5), there exists $a \in A$ such that $\mathrm{Ann}_R(a) = I$. Define $\phi\colon A \to A_M$ to be the natural homomorphism. $R_M\phi(a) \cong (R/I)_M \cong R_M/I_M$ and so $\mathrm{Ann}_{R_M}(\phi(a)) = I_M$. By 5.3(7) $P \cong P_M$ and so $I_M^c \cap R \subset I$. Thus $I = \mathrm{Ann}_R(a) \subset \mathrm{Ann}_R(\phi(a)) \subset I_M^c \cap R \subset I$, and hence $\mathrm{Ann}_R(\phi(a)) = I$. R is a Bezout ring and 3.7 implies R_M is a valuation ring. R_M is then an almost maximal valuation ring, so 4.2 implies R_M is an FGC ring. $A_M = \bigoplus_{i=1}^{n} R_M x_i$ for some $x_i \in A_M$. There exist $r_i \in R_M$ such that $\phi(a) = \sum_{i=1}^{n} r_i x_i$. $I = \mathrm{Ann}_R(\phi(a)) = \bigcap_{i=1}^{n} \mathrm{Ann}_R(r_i x_i)$. As above there exists an $i \in \{1,\dots,n\}$ such that $I = \mathrm{Ann}_R(r_i x_i)$. Let $\pi\colon A_M \to R_M x_i$ be the projection homomorphism for this i . $\pi\phi(A)$ is a finitely generated R-submodule of $R_M x_i \cong R_M/I_M$. R_M is a valuation ring and by getting common denominators, there exists $y \in R_M x_i$ such that $\pi\phi(A) = Ry$. $\mathrm{Ann}_R(y) = I$. Choose $b \in A$ such that $\pi\phi(b) = y$. Then one easily shows that $A = \ker(\pi\phi) \oplus Rb$.

Thus A has been decomposed into a direct sum of $\ker(\pi\phi)$ and a cyclic submodule Rb . One continues decomposing $\ker(\pi\phi)$ by repeating the above procedure. $\dim_{R/M}[\ker(\pi\phi)/M\ker(\pi\phi)] = \dim_{R/M}[A/MA] - 1$, so this process must stop after a finite number of steps. q.e.d.

At this point we have found three different types of FGC rings, namely almost maximal valuation rings, almost maximal Bezout domains, and torch rings. Each of these has a unique minimal prime ideal, so each is an indecomposable ring. By 4.1 a finite product of such rings is an FGC ring. Our main goal in part one is to prove that these are the only FGC rings. Consequently many of the following facts will start with the assumption that R is an FGC ring.

The construction used in the next proof originated with the work of R.S. Pierce [25], who was considering decompositions of finitely generated modules over commutative regular rings. R is said to be <u>regular</u> if for all $a \in R$ there exists $x \in R$ such that $a = axa$. S. Weigand [36] used a similar construction to show that in an FGC domain, every non-zero prime ideal is a subset of a unique maximal ideal. P. Vamos [31], and independently T. Shores (unpublished) gave the result as stated here. The proof given is that of T. Shores, and the S. Wiegand result will be an easy corollary.

<u>Theorem 5.5</u>: (P. Vamos [31]) If R is an FGC ring with unique minimal prime ideal P, then P is a uniserial R-module.

<u>Proof</u>: Suppose the result is false. Then there exist $x,y \in P$ such that Rx and Ry are not comparable. By Zorn's Lemma there exists an ideal I_1 of R, maximal among the ideals of R properly contained in Rx and containing $Rx \cap Ry$. Similarly there exists an ideal I_2 of R, maximal among the ideals of R properly contained in Ry and containing $Rx \cap Ry$. Replacing R by $R/I_1 + I_2$ and using 4.1(2) allows us to assume that $Rx \cong R/M$ and $Ry \cong R/M'$ for $M,M' \in \text{mspec}R$, and $Rx \cap Ry = \{0\}$. By 4.6 and 3.7 R_M is a valuation ring. $R_M x \neq \{0\}$ and $R_{M'} y \neq \{0\}$, and so $M \neq M'$. Replacing R by $R_{R-(M \cup M')}$ and using 4.1(3) allows us to assume that R is an FGC ring. $\text{mspec}R = \{M,M'\}$, $M \neq M'$, $\text{minspec}R = \{P\}$, $x,y \in P - \{0\}$, $Rx \cap Ry = \{0\}$, $Rx \cong R/M$, and $Ry \cong R/M'$. Thus Rx and Ry are simple ideals of R, and by localizing, these are the only simple ideals of R.

Let $I = \{r \in R\colon$ there exists $s \in R - M$ such that $rs = 0\}$. Thus I is the kernel of the canonical map $R \to R_M$. Similarly let $J = \{r \in R\colon$ there exists $s \in R - M'$ such that $rs = 0\}$ and so J is the kernel of $R \to R_{M'}$. Then $y \in I \subset P$, $x \in J \subset P$, $x \notin I$, and $y \notin J$.

Define $A = R/I \oplus R/J$ and use bars to denote modulo I or modulo J in the appropriate factor. The submodule $R(\bar{x},\bar{0})$ of A is isomorphic to $Rx \cong R/M$, and every non-zero submodule of the first factor of A must contain $R(\bar{x},\bar{0})$. Similarly $R(\bar{0},\bar{y}) \cong Ry \cong R/M'$ and every non-zero submodule of the second factor of A must contain $R(\bar{0},\bar{y})$. It follows that every non-zero submodule of A contains at least one of the simple submodules $R(\bar{x},\bar{0})$ or $R(\bar{0},\bar{y})$. In particular, these are the only two simple submodules of A .

$M + M' = R$ so there exist $u \in M - M'$ and $v \in M' - M$ such that $u + v = 1$. Define B to be submodule of A given by $B = R(\bar{1},\bar{u}) + R(\bar{0},\bar{v})$. The decomposition of B into a direct sum of cyclic submodules will lead us to the desired contradiction. We consider the local properties of B . $I_M = \{0\}$, $(R/I)_M \cong R_M$, and $R_M(\bar{1},\bar{u}) \cong R_M$. $R_M(\bar{0},\bar{v}) \cong (R/J)_M$. Thus $B_M \cong R_M(\bar{1},\bar{u}) \oplus R_M(\bar{0},\bar{v}) \cong R_M \oplus (R/J)_M$. Note that $B = R(\bar{1},\bar{1}) + R(\bar{v},\bar{0})$ and so as above $B_{M'} \cong R_{M'}(\bar{1},\bar{1}) \oplus R_{M'}(\bar{v},\bar{0}) \cong R_{M'} \oplus R_{M'}(\bar{v},\bar{0})$.

R is an FGC ring implies $B = B_1 \oplus \ldots \oplus B_k$ for cyclic submodules B_i of B . Since every non-zero submodule of A contains at least one of the two simple submodules of A , it must be the case that $k \leq 2$. That B is not cyclic can be seen by localizing and using the comments of the last paragraph. Thus $B = B_1 \oplus B_2$ with B_1 and B_2 non-zero cyclic submodules of B .

Since R_M is a valuation ring and the decomposition of finitely generated modules over valuation rings is unique, 3.4, we may assume that $(B_1)_M \cong R_M$, $(B_2)_M \cong (R/J)_M$, and either $(B_1)_{M'}$ or $(B_2)_{M'}$ is isomorphic to $R_{M'}$. Consider first the case where $(B_1)_{M'} \cong R_{M'}$. By 2.2 $B_1 \cong R$. But R has two simple submodules Rx and Ry , and so B_2 contains no simple submodules. This contradicts the statement that every non-zero submodule of A contains a simple submodule. The first case is not possible, so we assume the second case, i.e., $(B_2)_{M'} \cong R_{M'}$. $(B_2)_M = (R/J)_M$ and $(B_2)_{M'} \cong R_{M'} \cong (R/J)_{M'}$, and so by 2.2 $B_2 \cong R/J$. Suppose $B_2 = Rb$ with $b \in B_2$. Then $b = r(\bar{1},\bar{u}) + s(\bar{0},\bar{v})$ for some $r,s \in R$. If $r \in M'$ then $yb = 0$ and $y \in J$ which is a contradiction. Thus $r \notin M'$. $Jb = \{0\}$ implies $Jr = \{0\}$ and so $r \in M - M'$. In particular $r \notin P$. Thus $Rr \supset P$ by 2.2(1) . $x \in P \subset Rr$ implies $x = rt$ for some $t \in P$. $r^2t = rx = 0$, $t \in J$, $x = rt \in Jr = \{0\}$, contradicting $x \neq 0$. q.e.d.

Theorem 5.6: (S. Wiegand [36]) If R is an FGC domain, then every non-zero prime ideal of R is a subset of a unique maximal ideal of R .

Proof: Suppose the theorem is false and P is a non-zero prime ideal of the FGC domain R with $M,M' \in \text{mspec}(P)$, and $M \neq M'$. Replace R by $R_{R - (M \cup M')}$ and using 4.1(3) , we may assume $\text{mspec}R = \{M,M'\}$, $M \neq M'$, and $\{0\} \neq P \subset M \cap M'$. Choose $x \in P - \{0\}$, $m \in M - M'$, and $m' \in M' - M$. Then Rxm and Rxm' are non-comparable ideals of R contained in P . R_P is a local FGC ring by 4.1(3) and so R_P is a valuation ring by 4.3. Thus $\{J \in \text{spec}R\colon J \subset P\}$ is a chain. Let $P' = \bigcap\{J \in \text{spec}R\colon x \in J \subset P\}$. Then $P' \in \text{spec}R$. Let a bar denote modulo $Rxm \cap Rxm'$. $\bar{R}$ is an FGC ring by 4.1(2). $\overline{P'}$ is the unique minimal prime ideal of $\bar{R}$, and $\overline{Rxm}$ and $\overline{Rxm'}$ are non-comparable ideals of $\bar{R}$ contained in $\overline{P'}$. This contradicts 5.5 . q.e.d.

To prove the main theorem, the next fact needed is 8.5, i.e., that an FGC ring has only finitely many minimal prime ideals. This will require topological considerations. The next three sections give the necessary topological background.

Section 6 The Zariski and Patch Topologies of the Spectrum of a Ring

For a set X let $\underline{\mathcal{P}(X)}$ denote the set of all subsets of X. $\mathcal{P}(X)$ is refered to as the power set of X. For $I \in \mathcal{P}(R)$ define $\underline{D(I)} = \{P \in \text{specR}: I \not\subset P\}$ and $\underline{V(I)} = \{P \in \text{specR}: I \subset P\} = \text{specR} - D(I)$. For $x \in R$, $D(\{x\})$ will be shortened to $\underline{D(x)}$ and similarly for $\underline{V(x)}$. For $\{I_j\}_{j=1}^{n} \subset \mathcal{P}(R)$ we use the notation $I_1 I_2 \cdots I_n$ for the set all finite sums of terms of the form $x_1 x_2 \cdots x_n$ for $x_j \in I_j$. Thus for $I \in \mathcal{P}(R)$, RI is the smallest ideal of R containing I.

Lemma 6.1:

1. For $\{I_\alpha\}_{\alpha \in X} \subset \mathcal{P}(R)$, $\bigcup_{\alpha \in X} D(I_\alpha) = D(\bigcup_{\alpha \in X} I_\alpha)$.
2. For $\{I_j\}_{j=1}^{n} \subset \mathcal{P}(R)$, $\bigcap_{j=1}^{n} D(I_j) = D(\{x_1 x_2 \cdots x_n \in R: x_j \in I_j$ for $j = 1,\ldots,n\}) = D(I_1 I_2 \cdots I_n) = D(RI_1 I_2 \ldots I_n)$. In particular $D(I) = D(RI)$ for all $I \in \mathcal{P}(R)$.
3. For $\{I_\alpha\}_{\alpha \in X} \subset \mathcal{P}(R)$, $\bigcap_{\alpha \in X} V(I_\alpha) = V(\bigcup_{\alpha \in X} I_\alpha)$.
4. For $\{I_j\}_{j=1}^{n} \subset \mathcal{P}(R)$, $\bigcup_{j=1}^{n} V(I_j) = V(\{x_1 x_2 \ldots x_n \in R: x_j \in I_j$ for $j = 1,\ldots,n\}) = V(I_1 I_2 \ldots I_n) = V(RI_1 I_2 \ldots I_n)$. In particular $V(I) = V(RI)$ for all $I \in \mathcal{P}(R)$.

Proof:

1. $P \in \bigcup_{\alpha \in X} D(I_\alpha)$ if and only if for some $\alpha \in X$, $P \in D(I_\alpha)$ if and only if for some $\alpha \in X$, $I_\alpha \not\subset P$ if and only if $\bigcup_{\alpha \in X} I_\alpha \not\subset P$ if and only if $P \in D(\bigcup_{\alpha \in X} I_\alpha)$.
2. $P \in \bigcap_{j=1}^{n} D(I_j)$ if and only if for all j, $I_j \not\subset P$ if and only if for all j there exists $x_j \in I_j - P$ if and only if for all j there exists $x_j \in I_j - P$ with $x_1 x_2 \ldots x_n \notin P$ if and only if

$P \in D(\{x_1x_2\ldots x_n: \ x_j \in I_j \ \text{ for } \ j = 1,\ldots,n\})$. The other parts follow from the fact that the elements of specR are ideals of R .

3. and 4. The proofs are similar to parts 1 and 2 . q.e.d.

Definition: The Zariski topology of specR is the topology which has $\{D(I): \ I \in \mathcal{P}(R)\}$ as a subbasis of open sets.

By 6.1 one sees that $\{D(I): I \in \mathcal{P}(R)\} = \{D(I): I$ is an ideal of $R\}$ is the set of all open subsets of specR in the Zariski topology. Moreover $\{D(x): \ x \in R\}$ is a basis of open subsets of specR in the Zariski topology.

There is not much that can be said about the Zariski topology in general. For example consider the ring of integers, Z . The Zariski closed subsets of specZ are specZ and all finite subsets of specZ - {{0}} . Thus the Zariski topology of specZ is not Hausdorff, but it is compact. We shall see shortly that the Zariski topology is always compact.

We wish to define the patch topology of specR . Use 2 to denote the discrete topological space {0,1} . For a set X we denote the set of all functions $X \to 2$ by 2^X . Let $\chi\colon \ \mathcal{P}(X) \to 2^X$ be given by $\chi(A)$ is the characteristic function of A , for $A \in \mathcal{P}(X)$. Clearly χ is a set bijection.

Give 2^X the product topology. For $x \in X$, let $\pi_x\colon \ 2^X \to 2$ be the projection map onto the $x^{\underline{th}}$ factor, i.e., $\pi_x(f) = f(x)$ for $f \in 2^X$. Then a subbasis for the product topology of 2^X is given by

$\{\pi_x^{-1}(\{1\})\}_{x\in X} \cup \{\pi_x^{-1}(\{0\})\}_{x\in X}$. For $x \in X$

$\chi^{-1}(\pi_x^{-1}(\{1\})) = \{A \in \mathcal{P}(X): x \in A\}$ and $\chi^{-1}(\pi_x^{-1}(\{0\})) = \{A \in \mathcal{P}(X): x \notin A\}$.

Give $\mathcal{P}(X)$ the topology gotten from transfering the topology of 2^X by χ^{-1}. Thus a subbasis of this topology of $\mathcal{P}(X)$ is

$\{\{A \in \mathcal{P}(X): x \in A\}\}_{x\in X} \cup \{\{A \in \mathcal{P}(X): x \notin A\}\}_{x\in X}$.

Definition: The patch topology of specR is the subspace topology of of specR considered as a subset of $\mathcal{P}(R)$.

For $x \in R$, $\{A \in \mathcal{P}(R): x \in A\} \cap \text{specR} = V(x)$ and $\{A \in \mathcal{P}(R): x \notin A\} \cap \text{specR} = D(x)$. Thus a subbasis of the patch topology of specR is $\{V(x)\}_{x \in R} \cup \{D(x)\}_{x \in R}$. Using 6.1 and the fact that $D(1) = V(0) = \text{specR}$, a basis of the patch topology of specR is $\{D(a) \cap V(b_1) \cap \ldots \cap V(b_n): a, b_i \in R \text{ for } i = 1,\ldots,n\}$. Note that $\text{specR} - (D(a) \cap V(b_1) \cap \ldots \cap V(b_n)) = V(a) \cup D(b_1) \cup \ldots \cup D(b_n)$ which is a patch open subset of specR . Thus this basis of the patch topology consists of sets which are both open and closed in the patch topology.

<u>Lemma 6.2</u>: If R is a Bezout ring, then $\{D(a) \cap V(b): a,b \in R\}$ is a basis for the patch topology of specR , and the sets in this basis are both open and closed in the patch topology.

<u>Proof</u>: Suppose $b_1,\ldots,b_n \in R$. Since R is a Bezout ring, $\sum_{i=1}^{n} Rb_i = Rb$ for some $b \in R$. Then $V(b_1) \cap \ldots \cap V(b_n) = V(b)$. The lemma follows by the comments in the above paragraph. <u>q.e.d.</u>

<u>Lemma 6.3</u>: Every Zariski open subset of specR is patch open.

<u>Proof</u>: $\{D(x)\}_{x \in R}$ is a basis of the Zariski topology, and these sets are patch open. <u>q.e.d.</u>

Consider the example Z , the ring of integers. The family of all patch open subsets of specZ is $\{U \in \mathcal{P}(\text{specZ}): \{0\} \in U \text{ implies } |\text{specZ} - U| < \infty\}$ Thus in the patch topology, the subspace $\text{specZ} - \{\{0\}\}$ is discrete, and specZ is the one-point compactification of $\text{specZ} - \{\{0\}\}$.

M. Hochster [12] studied the patch topology of specR . He showed that specR is compact Hausdorff in the patch topology, this being the next result

<u>Definition</u>: A <u>Boolean space</u> is a topological space which is compact Hausdorff and totally disconnected.

<u>Theorem 6.4</u>: The patch topology makes specR a Boolean space.

<u>Proof</u>: 2^R is a Boolean space and the patch topology of specR is the

subspace topology of $\mathcal{P}(R) \cong 2^R$, so the patch topology of specR is Hausdorff and totally disconnected. To show compactness, it suffices to show that specR is a closed subset of $\mathcal{P}(R)$. To see this note that

$$\text{specR} = V'(0) \cap [\bigcap_{a,b\in R}(D'(a) \cup D'(b) \cup V'(a+b))] \cap [\bigcap_{a,r\in R}(D'(a) \cup V'(ra))] \cap [\bigcap_{a,b\in R}(V'(a) \cup V'(a) \cup D'(ab))]$$

where $D'(x) = \{A \in \mathcal{P}(R): x \notin A\}$ and $V'(x) = \{A \in \mathcal{P}(R): x \in A\}$ for all $x \in R$. This exhibits specR as an intersection of closed subsets of $\mathcal{P}(R)$ and so specR is a closed subset. q.e.d.

Corollary 6.5: specR is compact in the Zariski topology.

Proof: 6.3 and 6.4 .

Definitions: Let $Y \subset$ specR . The Zariski closure of Y in specR is denoted $\overline{Y}$. The patch closure of Y in specR is denoted Y^p . Y is a patch if $Y = Y^p$. Y is a thin patch if $Y = (\min Y)^p$, where the min operation is with respect to the partial ordering of set inclusion.

It is straight forward to see that if $Y \subset$ specR , then $\overline{Y} = V(\cap Y)$.

Theorem 6.6: (T. Shores and R. Wiegand [28]) Let Y be a patch of specR . Then the Zariski and patch subspace topologies of minY are the same.

Proof: Let X be a patch. We claim that $\overline{X} = \{P \in \text{specR}:$ there exists $P' \in X$ such that $P \supset P'\}$. The one inclusion is straight forward since $\overline{X} = V(\cap X)$. On the other hand, suppose $P \in \overline{X}$. Let $\mathcal{C} = \{D(x) \cap X: x \in R - P\}$ For $x \in R - P$, $x \notin \cap X$ and so $x \notin P_0$ for some $P_0 \in X$, i.e., $P_0 \in D(x) \cap X$. This shows that $\emptyset \notin \mathcal{C}$. P is a prime ideal and 6.1(2) implies $\mathcal{C}$ is closed under finite intersections. X is compact in the patch topology by 6.4 and $\mathcal{C}$ is a family of patch closed subsets of X with the f.i.p., so there exists $P' \in \cap\mathcal{C}$. Clearly $P' \in X$ and $P' \subset P$, showing that $\overline{X}$ has the required form.

Let Y be a patch of specR . Every Zariski closed subset of minY is a patch closed subset of minY by 6.3 . Suppose A is a patch closed subset of the space minY . It suffices to show that A is Zariski closed in the space minY . Let A' denote the Zariski closure of A in minY . Suppose $P \in A'$. Then $P \in \overline{A^p}$. By the first paragraph with $X = A^p$, there exists $P' \in A^p$ such that $P' \subset P$. $A^p \subset Y$ since Y is a patch, and so $P' \in Y$. Thus $P' \in Y$ and $P' \subset P$, hence $P' = P$ since $P \in \min Y$. $P = P' \in A^p \cap \min Y = A$ since A is patch closed in minY . This verifies that $A' = A$, and so A is a Zariski closed subset of the space minY . q.e.d.

Although we shall not use it in our development, we briefly remark how the topology of $\mathcal{P}(X)$ may be obtained by another method. For a net $\{A_\alpha\}_{\alpha \in D}$ of $\mathcal{P}(X)$ and $A \in \mathcal{P}(X)$ we say $\{A_\alpha\}_{\alpha \in D}$ converges to A if $A = \bigcap_{\alpha \in D} (\bigcup_{\beta \geq \alpha} A_\beta) = \bigcup_{\alpha \in D} (\bigcap_{\beta \geq \alpha} A_\beta)$. For $\mathcal{C} \subset \mathcal{P}(X)$, define $\mathcal{C}$ to be closed whenever $\{A_\alpha\}_{\alpha \in D}$ is a net of $\mathcal{C}$, $A \in \mathcal{P}(X)$, and $\{A_\alpha\}_{\alpha \in D}$ converges to A , then $A \in \mathcal{C}$. It can be shown that this family of closed subsets of $\mathcal{P}(X)$ is the family of closed sets of a topology of $\mathcal{P}(X)$, called the <u>order convergence topology</u>. It turns out that the order convergence topology of $\mathcal{P}(X)$ is the same as the topology of $\mathcal{P}(X)$ obtained in this section from 2^X .

Section 7 The Stone-Cech Compactification of N .

In this section we wish to define and verify some basic properties of βN , the Stone-Cech compactification of the discrete topological space N . As a set βN can be viewed as the set of all ultrafilters of N , hence the following preliminaries about ultrafilters.

A much used reference for the material on βN is the text by L. Gillman and M. Jerison [8] , although several recent texts also have extensive coverage of βN . One reference for the point set topology is the text by J. Dugundji [6] It contains all the necessary background material, including convergence in terms of filterbases.

Definition: Let X be a non-empty set. $\mathcal{F}$ is a filter of X if

1. $\mathcal{F} \subset \mathcal{P}(X) - \{\emptyset\}$,
2. $F_1 \cap F_2 \in \mathcal{F}$ for all $F_1, F_2 \in \mathcal{F}$, and
3. if $F \in \mathcal{F}$ and $F \subset F' \subset X$, then $F' \in \mathcal{F}$.

$\mathcal{F}$ is an ultrafilter of X if $\mathcal{F}$ is a filter of X and $\mathcal{F}$ is not properly contained in another filter of X .

Lemma 7.1: Let X be a non-empty set. Then:

1. Any filter of X is contained in an ultrafilter of X .
2. If $\mathcal{F}$ is an ultrafilter of X and $Y \subset X$, then $Y \in \mathcal{F}$ or $X - Y \in \mathcal{F}$.
3. If $\mathcal{F}$ is an ultrafilter of X and $X_1, X_2 \in \mathcal{P}(X) - \mathcal{F}$, then $X_1 \cup X_2 \in \mathcal{P}(X) - \mathcal{F}$.

Proof:

1. Use Zorn's Lemma.
2. Let $\mathcal{F}$ be an ultrafilter of X and $Y \subset X$. If there exist $F_1, F_2 \in \mathcal{F}$ such that $Y \supset F_1$ and $X - Y \supset F_2$, then $\emptyset = Y \cap (X - Y) \supset F_1 \cap F_2 \in \mathcal{F}$, contradicting $\emptyset \notin \mathcal{F}$. Thus without loss of generality, we may assume that $X - Y \not\supset F$ for all $F \in \mathcal{F}$. In other words $Y \cap F \neq \emptyset$ for all $F \in \mathcal{F}$. Let $\mathcal{G} = \{G \in \mathcal{P}(X) : Y \cap F \subset G$ for some $F \in \mathcal{F}\}$. $\mathcal{G}$ is a

filter of X and $\mathcal{F} \subset \mathcal{G}$. Since $\mathcal{F}$ is an ultrafilter of X, $\mathcal{F} = \mathcal{G}$. Then $Y \in \mathcal{G} = \mathcal{F}$.

3. If $\mathcal{F}$ is an ultrafilter of X and $X_1, X_2 \in \mathcal{P}(X) - \mathcal{F}$, then by part 2 $X - X_1, X - X_2 \in \mathcal{F}$, $X - (X_1 \cup X_2) = (X - X_1) \cap (X - X_2) \in \mathcal{F}$ and so $X_1 \cup X_2 \notin \mathcal{F}$. q.e.d.

Definition: Let $\mathcal{F}$ be a filter of X. Then $\mathcal{F}$ is a fixed filter if $\cap\mathcal{F} \neq \emptyset$, and $\mathcal{F}$ is a free filter if $\cap\mathcal{F} = \emptyset$.

For $x \in X$, let $\mathcal{F}(x) = \{A \in \mathcal{P}(X): x \in A\}$. Then $\mathcal{F}(x)$ is a fixed ultrafilter of X. Using 7.1, one sees that $\{\mathcal{F}(x)\}_{x \in X}$ is the set of all fixed ultrafilters of X. Suppose X is an infinite set and $\mathcal{F}_0 = \{A \in \mathcal{P}(x): |X - A| < \infty\}$. Then $\mathcal{F}_0$ is a filter of X. Let $\mathcal{F}$ be an ultrafilter of X containing $\mathcal{F}_0$. Then $\mathcal{F}$ is a free ultrafilter of X. If $\mathcal{F}$ is a free ultrafilter of X and $F \in \mathcal{F}$, then $|F| = \infty$. In particular, if X is a finite set, then the only ultrafilters on X are the fixed ultrafilters.

Definition: Let X be a non-empty set. Define βX to be the set of all ultrafilters of X. For $\mathcal{I} \subset \mathcal{P}(X)$, define $D(\mathcal{I}) = \{\mathcal{F} \in \beta X: \mathcal{I} \not\subset \mathcal{F}\}$ and $V(\mathcal{I}) = \{\mathcal{F} \in \beta X: \mathcal{I} \subset \mathcal{F}\} = \beta X - D(\mathcal{I})$. If $I \subset X$, then $D(\{I\})$ will be abbreviated $D(I)$ and similarly for $V(I)$. βX is topologized by taking $\{D(I): I \subset X\}$ as a subbasis of open sets of βX. Let $i_X: X \to \beta X$ be given by $i_X(x) = \mathcal{F}(x)$, the fixed ultrafilter at x, for $x \in X$.

Lemma 7.2: Let X be a non-empty set. Then:

1. If $\{I_\alpha\}_{\alpha \in Y} \subset \mathcal{P}(X)$, then $\bigcup_{\alpha \in Y} D(I_\alpha) = D(\{I_\alpha\}_{\alpha \in Y})$.
2. If $\{I_j\}_{j=1}^n \subset \mathcal{P}(X)$, then $\bigcap_{j=1}^n D(I_j) = D(\bigcup_{j=1}^n I_j)$.
3. If $\{I_\alpha\}_{\alpha \in Y} \subset \mathcal{P}(X)$, then $\bigcap_{\alpha \in Y} V(I_\alpha) = V(\{I_\alpha\}_{\alpha \in Y})$.
4. If $\{I_j\}_{j=1}^n \subset \mathcal{P}(X)$, then $\bigcup_{j=1}^n V(I_j) = V(\bigcup_{j=1}^n I_j)$.

Proof: Part 1 is a straight forward consequence of the definitions, and part 2 is also straight forward using 7.1(3). Parts 3 and 4 are comparable to parts 1 and 2. q.e.d.

We are leading up to the assertion that if X is a non-empty discrete topological space then $(\beta X, i_X)$ is the Stone-Cech compactification of X. By 7.2 $\{D(I): I \subset X\}$ is a basis of open sets of βX and $\{D(\mathcal{J}): \mathcal{J} \subset \mathcal{P}(X)\}$ is the set of all open subsets of βX. If $I \subset X$, then $\beta X - D(I) = D(X - I)$ so $D(I)$ is also a closed subset of βX.

Lemma 7.3: If X is a non-empty set, then βX is a Boolean space.

Proof: One could show directly from the definitions and lemmas of this section that βX is a compact Hausdorff and totally disconnected topological space. However, we prefer to proceed as follows. On $\mathcal{P}(X)$ we define the operations of addition and multiplication as follows:
for $A, B \in \mathcal{P}(X)$ let $A + B = X - ((A - B) \cup (B - A))$ and $A \cdot B = A \cup B$. Then $\mathcal{P}(X)$ is a ring, $\operatorname{spec}\mathcal{P}(X) = \beta X$, and the patch topology of $\operatorname{spec}\mathcal{P}(X)$ is the same as the topology of βX. By 6.4 $\operatorname{spec}\mathcal{P}(X)$ is a Boolean space in the patch topology, and so βX is a Boolean space. q.e.d.

We review some topological definitions and facts.

Definition: If X is a (completely regular) topological space, then (X', j) is a Stone-Cech compactification of X if

1. X' is a compact Hausdorff topological space,
2. $j: X \to X'$ is a continuous embedding of X onto a dense subset of X', and
3. if $f: X \to Y$ is a continuous function with Y a compact Hausdorff topological space, then there exists a unique continuous $\overline{f}: X' \to Y$ such that $f = \overline{f} \circ j$.

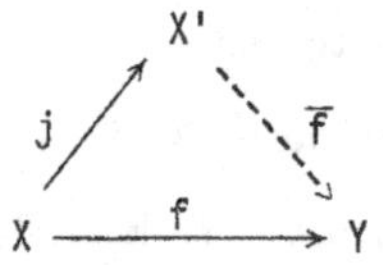

Actually the given definition is slightly redundant, since the uniqueness of $\overline{f}$ is equivalent to the denseness of $j(X)$ in X'. It is well known that a topological space X has a Stone-Cech compactification if and only if X is completely regular; and in that case the Stone-Cech compactification is unique in the obvious sense.

Definitions: If Y is a non-empty set, then $\mathcal{F}$ is said to be a filterbase of Y if $\emptyset \neq \mathcal{F} \subset \mathcal{P}(Y) - \{\emptyset\}$ and for all $F_1, F_2 \in \mathcal{F}$ there exists $F_3 \in \mathcal{F}$ such that $F_3 \subset F_1 \cap F_2$. $\mathcal{F}$ is said to be a maximal filterbase of Y if $\mathcal{F}$ is a filterbase of Y and for all $A \subset Y$ there exists $F \in \mathcal{F}$ such that $F \subset A$ or $F \subset Y - A$. A filterbase $\mathcal{F}$ of the topological space Y is said to converge to $y \in Y$ if for all neighborhoods U of y, there exists $F \in \mathcal{F}$ such that $F \subset U$.

Then a topological space Y is Hausdorff if and only if every converging filterbase of Y converges to exactly one point of Y. Also a topological space Y is compact if and only if every maximal filterbase of Y converges.

Proposition 7.4: If X is a non-empty discrete topological space, then $(\beta X, i_X)$ is a Stone-Cech compactification of X.

Proof: By 7.3 βX is compact Hausdorff. Since X is discrete, i_X is continuous. Let $D(I)$ be a non-empty basic open subset of βX with $I \subset X$. One must have $I \neq X$ since $X \in \mathcal{F}$ for all $\mathcal{F} \in \beta X$. If $x \in X - I$, then $i_X(x) = \mathcal{F}(x) \in i_X(X) \cap D(I)$, showing $i_X(X)$ is dense in $\beta(X)$. Clearly i_X is one-to-one.

Suppose Y is a compact Hausdorff topological space and $f: X \to Y$ is continuous. We must define a function $\overline{f}: \beta X \to Y$. Let $\mathcal{F} \in \beta X$. $\{f(F)\}_{F \in \mathcal{F}}$ is a maximal filterbase of Y. Since Y is compact, there exists $y \in Y$ such that $\{f(F)\}_{F \in \mathcal{F}}$ converges to y. Define $\overline{f}(\mathcal{F}) = y$. This defines the function $\overline{f}$. We first verify that $f = \overline{f} \circ i_X$. If $x \in X$, then $\{f(F)\}_{F \in \mathcal{F}(x)}$ converges to $f(x)$, and since Y is Hausdorff, this filterbase converges only to $f(x)$. Thus $\overline{f}(i_X(x)) = \overline{f}(\mathcal{F}(x)) = f(x)$, showing $f = \overline{f} \circ i_X$

We verify that $\overline{f}$ is continuous. Suppose $\mathcal{F} \in \beta(X)$, $\overline{f}(\mathcal{F}) = y$, and U is an open neighborhood of y in Y . Since Y is compact Hausdorff, there exists an open neighborhood W of y such that $W \subset \overline{W} \subset U$. We claim that $V(f^{-1}(W))$ is an open neighborhood of $\mathcal{F}$ in βX and $\overline{f}(V(f^{-1}(w))) \subset U$. W is a neighborhood of y and $\{f(F)\}_{F \in \mathcal{F}}$ converges to y implies there exists $F_0 \in \mathcal{F}$ such that $f(F_0) \subset W$. $F_0 \subset f^{-1}(f(F_0)) \subset f^{-1}(W)$ and so $\mathcal{F} \in V(f^{-1}(W)) = D(X - f^{-1}(W))$. Thus $V(f^{-1}(W))$ is an open neighborhood of $\mathcal{F}$ in βX . Suppose $\mathcal{G} \in V(f^{-1}(W))$. Since $f(f^{-1}(W)) \subset W$, $f^{-1}(W) \in \mathcal{G}$, and $\{f(G)\}_{G \in \mathcal{G}}$ converges to $\overline{f}(\mathcal{G})$, we have $\overline{f}(\mathcal{G}) \in \overline{W} \subset U$. Hence $\overline{f}(V(f^{-1}(W))) \subset U$, and this shows that $\overline{f}$ is continuous. The function $\overline{f}$ is unique since $i_X(X)$ is dense in βX . This verifies all the conditions needed to show that $(\beta X, i_X)$ is a Stone-Cech conpactification of X . <u>q.e.d.</u>

From now on, given a discrete topological space X we will identify X with the subset $i_X(X)$ of βX . N will always be assumed to have the discrete topology. Thus we consider $N \subset \beta N$. Note that $\beta X - X$ is a closed subset of βX and X is an open subset of βX since $X = \bigcup_{x \in X} D(X - \{x\})$. The following theorem is well known, and appears in the L. Gillman and M. Jerison text [8] .

<u>Lemma 7.5</u>: Let X be a non-empty discrete topological space. If U is an open subset of βX , then $U \subset \overline{U \cap X}$.

<u>Proof</u>: This is just a special case of the topological fact that if D is a dense subset of the topological space Y and V is an open subset of Y , then $V \subset \overline{V \cap D}$. <u>q.e.d.</u>

<u>Theorem 7.6</u>: Let X be a non-empty discrete topological space. If C is an infinite closed subset of βX, then C contains a closed subset homeomorphic to βN . If C is an infinite closed subset of βX , then C contains a closed subset homeomorphic to $\beta N - N$.

Proof: We have remarked that $\beta N - N$ is a closed subset of βN, so the last statement of the theorem is a consequence of the other statement.

If $|C \cap X| = \infty$, choose a countably infinite subset C' of $C \cap X$. Then $\overline{C'}$ is homeomorphic to βN. For to see this, let $g: C' \to N$ be a set bijection. Define $\overline{g}: \overline{C'} \to \beta N$ by $\overline{g}(\mathcal{F}) = \{g(F \cap C'): F \in \mathcal{F}\}$ for $\mathcal{F} \in \overline{C'}$. Then $\overline{g}$ is the required homeomorphism. Thus we may assume $|C \cap X| < \infty$. Since $\beta X - X$ is a closed subset of βX, we may assume $C \subset \beta X - X$.

Thus assume C is an infinite closed subset of $\beta X - X$ and it suffices to show that C contains a closed subset homeomorphic to βN. βX is Hausdorff and has a basis consisting of open and closed subsets implies there exists $\{U_n\}_{n \in N}$ a family of pairwise disjoint open subsets of βX such that $U_n \cap C \neq \emptyset$ for all $n \in N$. Choose $y_n \in U_n \cap C$ and let $Y = \{y_n\}_{n \in N}$. Y is a discrete subspace of C and $\overline{Y} \subset C$. We will show that $\overline{Y}$ is homeomorphic to βN by showing that $(\overline{Y}$, inclusion$)$ is a Stone-Cech compactification of Y.

Let $g: Y \to H$ be continuous with H compact Hausdorff. Define $f: X \to H$ by $f(x) = g(y_n)$ if $x \in X \cap U_n$ and $f(x) = h_0$ if $x \in X - \bigcup_{n \in N} U_n$ where h_0 is any fixed element of H. f is continuous since X is discrete. By the universal property of βX, there exists a continuous $F: \beta X \to H$ such that $F|X = f$. For $n \in N$, $X \cap U_n \subset F^{-1}(g(y_n))$ and $F^{-1}(g(y_n))$ is closed, so $\overline{X \cap U_n} \subset F^{-1}(g(y_n))$. By 7.5, $U_n \subset F^{-1}(g(y_n))$. This verifies that $F|Y = g$. Thus given $g: Y \to H$ with H compact Hausdorff, there exists a continuous $F|\overline{Y}: \overline{Y} \to H$ such that $(F|\overline{Y})|Y = g$. By denseness, this extension of g is unique. It follows that $(\overline{Y}$, inclusion$)$ is a Stone-Cech compactification of Y. Y is homeomorphic to N, so by the uniqueness of the Stone-Cech compactification, $\overline{Y}$ is homeomorphic to βN. q.e.d.

Definition: Let α be a cardinal number, X a topological space, and $x \in X$. Then x is an α-point of X if there exists α pairwise disjoint open subsets of X such that x is an element of the closure of each.

If $\alpha > \gamma$ are cardinal numbers and x is an α-point of X, then x is a γ-point of X. In the next section we will need to know that $\beta N - N$ has a 3-point. This was first proved by R.S. Pierce [25], but only by assuming the continuum hypothesis. The next theorem by N. Hindman [10] is an improvement, since this shows that $\beta N - N$ has a 3-point without using the continuum hypothesis. The proof presented here is taken directly from the original source. The preliminary lemma is a special case of a result published in 1928 by A. Tarski [29]. The cardinality of the set of all real numbers is denoted $\underline{c}$, which of course is also the cardinality of $\mathcal{P}(N)$.

Lemma 7.7: There exists a family $\mathcal{A}$ of subsets of N such that $|\mathcal{A}| = c$, $|A| = \infty$ for all $A \in \mathcal{A}$, and $|A_1 \cap A_2| < \infty$ for all distinct $A_1, A_2 \in \mathcal{A}$.

Proof: Let $\{B_n\}_{n \in N}$ be a partition of N such that $|B_n| = \infty$ for all $n \in N$. For all $n_1 \in N$ let $\{B_{n_1 n}\}_{n \in N}$ be a partition of B_{n_1} such that $|B_{n_1 n}| = \infty$ for all $n \in N$. Inductively, after $B_{n_1 n_2 \ldots n_k}$ have been defined for all $n_1, n_2, \ldots, n_k \in N$, let $\{B_{n_1 n_2 \ldots n_k n}\}_{n \in N}$ be a partition of $B_{n_1 n_2 \ldots n_k}$ such that $|B_{n_1 n_2 \ldots n_k n}| = \infty$ for all $n \in N$.

Let $\mathcal{S}$ be the set of all sequences $\{n_k\}_{k \in N}$ where $n_k \in N$ for all $k \in N$. For $S = \{n_k\}_{k \in N} \in \mathcal{S}$ define B_S as follows:

choose $x_1 \in B_{n_1}$, $x_2 \in B_{n_1 n_2} - \{x_1\}, \ldots, x_k \in B_{n_1 n_2 \ldots n_k} - \{x_1, \ldots, x_{k-1}\}, \ldots$, and let $B_S = \{x_k\}_{k \in N}$. Then $\mathcal{A} = \{B_S\}_{S \in \mathcal{S}}$ is the required family. q.e.d.

Theorem 7.8: (N. Hindman [10]) There exists a c-point in $\beta N - N$.

Proof: Let $\mathcal{A}$ be the family of subsets of N obtained in 7.7. For each $A \in \mathcal{A}$ choose $x_A \in V(A) - N$. If $A_1 \neq A_2$, then $x_{A_1} \neq x_{A_2}$ since $|A_1 \cap A_2| < \infty$ and no ultrafilter of $\beta N - N$ contains a finite subset of N.

Let $B = \{x_A\}_{A \in \mathscr{A}}$. Then $|B| = |\mathscr{A}| = c$.

We claim that there exists $y \in \beta N - N$ such that every neighborhood U of y has the property $|U \cap B| = c$. For suppose not. For all $x \in \beta N - N$ find an open neighborhood U_x of x such that $|U_x \cap B| < c$. $\{U_x \cap (\beta N - N)\}_{x \in \beta N - N}$ is an open cover of $\beta N - N$. $\beta N - N$ is compact, so there exists a finite subcover, and one deduces $|B| < c$, which is a contradiction. This verifies that there exists $y \in \beta N - N$ such that every neighborhood U of y has the property $|U \cap B| = c$. We will show that this y is the required c-point of $\beta N - N$.

y is a free ultrafilter of N so write $y = \{Y_\alpha\}_{\alpha < c}$, where $\alpha < c$ indicates that the α ranges over all ordinals less than the first ordinal of cardinality c . We wish to construct $\{X_\alpha\}_{\alpha < c}$ such that $|X_\alpha| = \infty$ and $X_\alpha \subset Y_\alpha$ for all $\alpha < c$, and $|X_\alpha \cap X_{\alpha'}| < \infty$ for all $\alpha \neq \alpha'$. For this purpose we construct an appropriate $\{A_\alpha\}_{\alpha < c} \subset \mathscr{A}$ by transfinite induction. $V(Y_1)$ contains c elements of B , so choose $A_1 \in \mathscr{A}$ such that $x_{A_1} \in V(Y_1)$. Let $\gamma < c$ and assume $A_\alpha \in \mathscr{A}$ have been chosen for all $\alpha < \gamma$ such that $|Y_\alpha \cap A_\alpha| = \infty$ and $A_\alpha \neq A_{\alpha'}$ for $\alpha \neq \alpha'$. $V(Y_\gamma)$ contains c elements of B , so choose $A_\gamma \in \mathscr{A}$ such that $A_\gamma \neq A_\alpha$ for all $\alpha < \gamma$ and $x_{A_\gamma} \in V(Y_\gamma)$. Then define $X_\alpha = Y_\alpha \cap A_\alpha$ for all $\alpha < \gamma$. This completes the construction of $\{X_\alpha\}_{\alpha < c}$ such that $|X_\alpha| = \infty$ and $X_\alpha \subset Y_\alpha$ for all $\alpha < c$ and $|X_\alpha \cap X_{\alpha'}| < \infty$ for $\alpha \neq \alpha'$.

For each $\alpha < c$ apply 7.7 to the set X_α to find a family $\{X_{\alpha\gamma}\}_{\gamma < c}$ such that $|X_{\alpha\gamma}| = \infty$ and $X_{\alpha\gamma} \subset X_\alpha$ for all $\gamma < c$, and $|X_{\alpha\gamma} \cap X_{\alpha\gamma'}| < \infty$ for $\gamma \neq \gamma'$. For each $\gamma < c$ define $U_\gamma = \bigcup_{\alpha < c} V(X_{\alpha\gamma}) \cap (\beta N - N)$. We claim that $\{U_\gamma\}_{\gamma < c}$ is a family of pairwise disjoint open subsets of $\beta N - N$ with $y \in \overline{U}_\gamma$ for all $\gamma < c$. The fact that $|X_{\alpha\gamma} \cap X_{\alpha'\gamma'}| < \infty$ for $(\alpha,\gamma) \neq (\alpha',\gamma')$ implies that the U_γ's are pairwise disjoint. Clearly U_γ

is an open subset of $\beta N - N$. Let $\gamma < c$ and let $D(I)$ be a basic open neighborhood of y with $I \subset N$. Then $N - I \in y$, so $N - I = Y_\alpha$ for some $\alpha < c$. $X_{\alpha\gamma} \subset X_\alpha \subset Y_\alpha = N - I$. Since $|X_{\alpha\gamma}| = \infty$, there exists $p \in \beta N - N$ such that $X_{\alpha\gamma} \in p$. Thus $p \in D(I) \cap U_\gamma$, $y \in \overline{U}_\gamma$, and so y is the required c-point of $\beta N - N$. <u>q.e.d.</u>

<u>Corollary 7.9</u>: $c \leq |\beta N| \leq 2^c$.

<u>Proof</u>: By 7.8, $c \leq |\beta N|$. $\beta N \subset \mathcal{P}(\mathcal{P}(N))$, so $|\beta N| \leq |\mathcal{P}(\mathcal{P}(N))| = 2^c$. <u>q.e.d.</u>

7.9 is incomplete in the sense that it is well known that $|\beta N| = 2^c$. For example see 9.2 of L. Gillman and M. Jerison [8] or p.244 of J. Dugundji [6]. For our purposes, we will only need to know that $c \leq |\beta N|$.

That $\beta N - N$ has a 3-point, 7.8, will be of importance for us in the next section. It is perhaps surprising that the proof actually produces a c-point. If one tries to show that $\beta N - N$ has a 3-point directly, one might start considering arbitrary infinite compact Hausdorff spaces. In this endeavor, we probably would notice that the subspace $\{0\} \cup \{1/n: n \in N\}$ of the real line has 0 as a 3-point. Similarly, if an infinite compact Hausdorff space has a converging sequence consisting of distinct points, then the limit of this sequence is a 3-point. In particular, any infinite compact metric space has a 3-point. One might be led to conjecture that every infinite compact Hausdorff space has a 3-point. However, this is not true since R.S. Pierce [25] shows that βN has no 3-point.

Section 8 Relating Topology to the Decomposition of Modules

While studying commutative regular rings, R.S. Pierce [25] showed that the decomposition of finitely generated modules is related to the spectrum of the ring with the Zariski topology. Of major importance in proving the main theorem is the work of T. Shores and R. Wiegand [28], who extended these ideas of R.S. Pierce by relating the property of R being an FGC ring to topological properties of $\operatorname{spec} R$. A result of R.S. Pierce and these topological considerations were used by W. Brandal and R. Wiegand [5] to show that an FGC ring can have only finitely many minimal prime ideals. This will be used in the next section to decompose FGC rings into a finite direct sum of indecomposable FGC rings of the type that have been discussed. All the proofs of the theorems in this section are taken from their original sources.

<u>Theorem 8.1</u>: (T. Shores and R. Wiegand [28]) If $\operatorname{spec} R$ contains pairwise disjoint Zariski open sets U_1, U_2, and U_3 such that $U_1^P \cap U_2^P \cap U_3^P \neq \emptyset$, then R is not an FGC ring.

<u>Proof</u>: Assume $\operatorname{spec} R$ contains pairwise disjoint Zariski open sets U_1, U_2, and U_3 with $P \in U_1^P \cap U_2^P \cap U_3^P$. $U_i = D(I_i)$ where I_i is an ideal of R, for $i = 1,2,3$. Define the R-module A by $A = R^2/T$ where $T = \{(r_1 + r_3, r_2 + r_3) \in R^2 : r_i \in I_i \text{ for } i = 1,2,3\}$. A is a finitely generated R-module, and it will be shown that A is not a direct sum of cyclic submodules.

For $M \in \operatorname{spec} R$ we use the notation $R(M)$ for the field R_M/MR_M and $A(M)$ for the $R(M)$-vector space A_M/MA_M; and for $r \in R$ and $a \in A$ let $r(M)$ and $a(M)$ denote the images of r and a in $R(M)$ and $A(M)$ respectively. b_1 and b_2 will denote $(1,0) + T$ and $(0,1) + T$, the elements of A.

Since $P \in U_2^P$ it follows that $P \notin U_1 = D(I_1)$, so $I_1 \subset P$. Similarly $I_2 \subset P$ and $I_3 \subset P$. For $i = 1,2,3$ define $E_i = \{a/s + PA_P \in A(P): s \in R - P, a \in A, \text{ and } a(M) = 0 \text{ for all } M \in U_i\}$. E_i is an $R(P)$-subspace of $A(P)$.

We claim that if $a(P) \in E_i$ for $a \in A$, then there exists $M \in U_i$ and $s \in R - M$ such that $sa \in MA$. For $a(P) \in E_i$ implies there exist $a' \in A$ and $t \in R - P$ such that $at - a' \in PA$ and $a'(M) = 0$ for all $M \in U_i$. Thus $at = a' + p_1b_1 + p_2b_2$ with $p_1, p_2 \in P$. Let $U = D(t) \cap V(p_1) \cap V(p_2)$. Then U is a patch open set and $P \in U_i^{\,p} \cap U$. Hence there exists $M \in U_i \cap U$. Then $a(M)t(M) = a'(M) + p_1(M)b_1(M) + p_2(M)b_2(M) = 0$. $t(M) \neq 0$ implies $a(M) = 0$, and the desired claim follows.

We claim that $b_1(P) \in E_1$. Let $M \in U_1 = D(I_1)$, and choose $r_1 \in I_1 - M$. Then $r_1b_1 = (r_1,0) + T = 0$ and so $r_1(M)b_1(M) = 0$. But $r_1(M) \neq 0$, so $b_1(M) = 0$. This verifies the claim that $b_1(P) \in E_1$. We claim that $b_2(P) \notin E_1$ For suppose $b_2(P) \in E_1$. By the last paragraph, there exist $M \in U_1$ and $s \in R - M$ such that $sb_2 \in MA$. Thus $sb_2 = m_1b_1 + m_2b_2$ for some $m_1, m_2 \in M$, $(m_1, m_2 - s) \in T$, $(m_1, m_2 - s) = (r_1 + r_3, r_2 + r_3)$ for some $r_i \in I_i$. Then $s = m_2 - r_2 - r_3 \in M + I_2 + I_3$. $M \in U_1$ implies $M \notin U_2 \cup U_3$, $I_2 + I_3 \subset M$. Hence $s \in M$, a contradiction. This verifies the claim that $b_2(P) \notin E_1$. In a similar manner one can verify that $b_1(P) \notin E_2$, $b_2(P) \in E_2$, $(b_1 + b_2)(P) \in E_3$, and $b_1(P) \notin E_3$. $A(P)$ is clearly a two dimensional $R(P)$-vector dimensional $R(P)$-vector space, so these conditions imply that the E_i's are distinct one-dimensional $R(P)$-subspaces of $A(P)$.

Suppose A is a direct sum of cyclic submodules, say $A = \bigoplus_{j=1}^{k} Ra_j$ for $a_j \in A$. Then $A(P) = \bigoplus_{j=1}^{k} R(P)a_j(P)$, and $A(P)$ being two dimensional, with a possible relabelling we may assume that $\{a_1(P), a_2(P)\}$ is a basis of $A(P)$. For a fixed i , choose a generator of E_i of the form $a(P)$ where $a \in A$ with $a(M) = 0$ for all $M \in U_i$. $a = \sum_{j=1}^{k} r_ja_j$ for some $r_j \in R$. For all $M \in U_i$ and all j , $r_ja_j(M) = 0$, and so $r_1(P)a_1(P)$, $r_2(P)a_2(P) \in E_1$. Since $r_1(P)$ and $r_2(P)$ cannot both be zero, it follows that $a_1(P) \in E_i$

or $a_2(P) \in E_i$. This being true for all $i = 1,2,3$, contradicts the previous paragraph that the E_i's are distinct one dimensional subspaces of $A(P)$. This contradiction verifies that A is not a direct sum of cyclic submodules, and so R is not an FGC ring. q.e.d.

The above theorem is not convenient to apply since two different topologies are involved. Better is the following result since it uses only one topology.

Theorem 8.2: (T. Shores and R. Wiegand [28]) If specR contains a thin patch with a 3-point relative to the patch topology, then R is not an FGC ring.

Proof: Suppose Y is a thin patch of specR with a 3-point relative to the patch topology. Let P be a 3-point of Y and let U_1, U_2, U_3 be pairwise disjoint patch open subsets of Y such that $P \in U_1^p \cap U_2^p \cap U_3^p$. By 6.6 there exist Zariski open subsets W_i of specR such that $W_i \cap \min Y = U_i \cap \min Y$ for $i = 1,2,3$. Y is a thin patch implies $\min Y$ is Zariski dense in $\overline{Y}$, and so $W_1 \cap \overline{Y}$, $W_2 \cap \overline{Y}$, and $W_3 \cap \overline{Y}$ are pairwise disjoint. $P \in (W_i \cap \overline{Y})^p$ for all $i = 1,2,3$. Relative to both the Zariski and the patch topologies, $\overline{Y}$ is homeomorphic to $\mathrm{spec}(R/\cap Y)$. By 8.1 $R/\cap Y$ is not an FGC ring, and so R is not an FGC ring. q.e.d.

Theorem 8.3: (R.S. Pierce [25]) If X is a Boolean space with a countably infinite dense subset consisting of isolated points of X , then either X contains a 3-point or X has a subspace homeomorphic to $\beta N - N$.

Proof: Suppose X is a Boolean space and Y is a countably infinite dense subset of X consisting of isolated points of X . We consider N to have the discrete topology and so there exists a homeomorphism $\phi_0 \colon N \to Y$. (βN, inclusion) is a Stone-Cech compactification of N by 7.4, so there exists a continuous $\phi \colon \beta N \to X$ such that $\phi|N = \phi_0$. Clearly $\phi(\beta N - N) = X - Y$.

Assume that there exists $x \in X$ such that $|\phi^{-1}\{x\}| \geq 3$. Choose distinct $p_1,p_2,p_3 \in \phi^{-1}\{x\}$, and find pairwise disjoint open neighborhoods U_1,U_2,U_3 of p_1,p_2,p_3 respectively. Then $p_i \in \overline{U_i \cap N}$ by 7.5, and $x = \phi(p_i) \in \phi(\overline{U_i \cap N}) \subset \overline{\phi(U_i \cap N)}$ for all $i = 1,2,3$. $\phi(U_i \cap N) \subset Y$ implies $\phi(U_i \cap N)$ is open in X, and the $\phi(U_i \cap N)$'s are pairwise disjoint. Thus x is a 3-point of X.

It suffices to consider the case that $|\phi^{-1}\{x\}| \leq 2$ for all $x \in X$. Define $K = \{x \in \beta N: |\phi^{-1}\{\phi(x)\}| = 1\}$ and $L = \{x \in \beta N: |\phi^{-1}\{\phi(x)\}| = 2\}$. Then $\{K,L\}$ is a partition of βN and $N \subset K$. If $K - N$ contains an infinite closed subset of $\beta N - N$, then by 7.6 $K - N$ and hence X contains a subspace homeomorphic to $\beta N - N$. Thus it suffices to consider the case where $K - N$ does not contain an infinite closed subset of $\beta N - N$. Then L is dense in $\beta N - N$, since every non-empty open subset of $\beta N - N$ contains a non-empty basic open set of the form $D(A) \cap (\beta N - N)$ for $A \subset N$, and such a set is infinite and closed.

For each $x \in L$ define $x' \in L$ by the condition $\phi^{-1}\{\phi(x)\} = \{x,x'\}$. Choose $x_1 \in L$. $x_1 \neq x_1' \in L$ and so there exist infinite sets A_1, A_2, A_3 such that $\{A_1, A_2, A_3\}$ is a partition of N, $x_1 \in V(A_1)$, $x_1' \in V(A_2)$, $x_1 \notin V(A_3)$, and $x_1' \notin V(A_3)$. Let $U_1 = V(A_1)$ and $V_1 = V(A_2)$. Then U_1 and V_1 are disjoint open and closed neighborhoods of x_1 and x_1' respectively and $U_1 \cup V_1 \not\supset \beta N - N$. Suppose we have found $\{x_1,\ldots,x_n\} \subset L$ with $\phi(x_i) \neq \phi(x_j)$ for $i \neq j$ and disjoint open and closed subsets U_n and V_n of βN (each of the form $V(A)$ for some $A \subset N$) such that $U_n \supset \{x_1,\ldots,x_n\}$, $V_n = \{x_1',\ldots,x_n'\}$, and $U_n \cup V_n \not\supset \beta N - N$. Since L is dense in $\beta N - N$, there exists $y \in L - (U_n \cup V_n)$. Then $\phi(y) \notin \{\phi(x_1),\ldots,\phi(x_n)\}$. There are three mutually exclusive possibilities: $y' \in U_n$, $y' \in V_n$, or $y' \notin U_n \cup V_n$. In the first possibility let $x_{n+1} = y'$ and in the other two possibilities let $x_{n+1} = y$. In each case one can find disjoint open and closed subsets U_{n+1} and V_{n+1} of βN (each of the form $V(A)$ for some $A \subset N$) such that $x_{n+1} \in U_{n+1} \supset U_n$,

$x'_{n+1} \in V_{n+1} \supset V_n$, and $U_{n+1} \cup V_{n+1} \not\supset \beta N - N$. Thus recursively we have constructed $\{x_n\}_{n\in N} \subset L$ and $\{U_n\}_{n\in N} \cup \{V_n\}_{n\in N}$ a family of open and closed subsets of βN (each of the form $V(A)$ for some $A \subset N$) such that $\phi(x_i) \neq \phi(x_j)$ if $i \neq j$, $U_n \cap V_n = \emptyset$, $U_{n+1} \supset U_n \supset \{x_1,\ldots,x_n\}$, and $V_{n+1} \supset V_n \supset \{x_1',\ldots,x_n'\}$. Let $U = \bigcup_{n\in N} U_n$ and $V = \bigcup_{n\in N} V_n$. Then $\overline{U} \cap \overline{V} = \emptyset$ For $n \in N$, $\phi(x_n) \in \phi(U) \subset \phi(\overline{U})$ and $\phi(x_n) = \phi(x_n') \in \phi(\overline{V})$. Let $W = \phi(\overline{U}) \cap \phi(\overline{V})$. By compactness, W is an infinite closed subset of X .

Consider $w \in W$. Then $w \in \phi(\overline{U})$, $\phi^{-1}\{w\} \cap \overline{U} \neq \emptyset$ and similarly $\phi^{-1}\{w\} \cap \overline{V} \neq \emptyset$. Thus for each $w \in W$, $\phi^{-1}\{w\}$ consists of exactly one point of $\overline{U}$ and one point of $\overline{V}$. Consequently $\phi|\overline{U} \cap \phi^{-1}W$ maps an infinite closed subset of $\beta N - N$ one-to-one onto W . By compactness this restriction of ϕ is a homeomorphism. By 7.6 W and hence X contains a subspace homeomorphic to $\beta N - N$. <u>q.e.d.</u>

<u>Lemma 8.4</u>: If X is an infinite Hausdorff topological space, then there exists a countably infinite discrete subspace of X .

<u>Proof</u>: Let $V_1 = X$. Let x_1, x_2 be distinct elements of X . There exists an open neighborhood U of x_1 such that $x_2 \notin \overline{U}$. Then either U or $X - U$ is infinite. Thus there exists $y_1 \in V_1$ and an infinite open subset V_2 of V_1 , such that $y_1 \notin \overline{V_2}$. Similarly there exists $y_2 \in V_2$ and an infinite open subset V_3 of V_2 such that $y_2 \notin \overline{V_3}$. Recursively one constructs $\{y_n\}_{n\in N} \subset X$ and $\{V_n\}_{n\in N}$ a family of infinite open subsets of X such that $V_n \supset V_{n+1}$ and $y_n \in V_n - \overline{V}_{n+1}$ for all $n \in N$. Then $\{y_n\}_{n\in N}$ is the required countably infinite discrete subspace of X . <u>q.e.d.</u>

<u>Theorem 8.5</u>: (W. Brandal and R. Wiegand [5]) If R is an FGC ring, then minspecR is finite.

<u>Proof</u>: Suppose the result is not true, and R is an FGC ring with minspecR infinite. In this proof all the topological properties of specR will be

relative to the patch topology. By 6.4 minspecR is Hausdorff so by 8.4 there exists $Y \subset$ minspecR such that Y is countably infinite and discrete. Thus the elements of Y are isolated points of Y^p. Y^p is a thin patch so by 8.2 Y^p does not contain a 3-point. By 8.3 Y^p contains a subspace V_1 homeomorphic to $\beta N - N$. By 7.8 and 8.2 V_1 is not a thin patch. The patch topology has a basis consisting of open and closed sets, so there exists a non-empty open and closed subset U of V_1 such that $U \cap \min V_1 = \emptyset$. U is infinite since every non-empty open subset of $\beta N - N$ is infinite. By 7.6 there exists a subset V_2 of U such that V_2 is homeomorphic to $\beta N - N$, and so $V_2 \cap \min V_1 = \emptyset$. Recursively one constructs a family $\{V_n\}_{n\in N}$ such that $Y^p \supset V_1 \supset V_2 \supset \ldots \supset V_n \supset \ldots$, $V_{n+1} \cap \min V_n = \emptyset$, and V_n is homeomorphic to $\beta N - N$ for all $n \in N$. Since $\beta N - N$ is compact, the V_n's are compact and closed. By the compactness of V_1, there exists $M \in \bigcap_{n\in N} V_n$.

R_M is a valuation ring by 4.6 and 3.7. Hence $\{P \in \operatorname{spec}R: P \subset M\}$ is a chain with respect to set inclusion. $\{P \in \operatorname{spec}R: P \subset M\} = \bigcap_{x\in R-M} D(x)$ and so this is a closed set. Thus $V_n \cap \{P \in \operatorname{spec}R: P \subset M\} = \{P \in V_n: P \subset M\}$ is a closed set. Define $P_n = \bigcap \{P \in V_n: P \subset M\}$. Then $P_n \in \operatorname{spec}R$, and since $\{P \in V_n: P \subset M\}$ is a closed set, $P_n \in V_n$. Clearly $P_n \in \min V_n$. It follows that $P_1 \subsetneq P_2 \subsetneq \ldots \subsetneq P_n \ldots$. Let $W = \{P_n\}_{n\in N} \cup \{\bigcup_{n\in N} P_n\}$. W is a closed set. $W \subset V_1$ and V_1 is homeomorphic to $\beta N - N$, so by 7.6 W contains a subset homeomorphic to βN. $c \leq |\beta N|$ by 7.9, so $c \leq |\beta N| \leq |W| = |N|$, contradiction. <u>q.e.d.</u>

It should be noted that all the topological considerations in sections 6,7, and 8 were used to prove 8.5, and 8.5 is the only result in these three sections that is needed later.

Section 9 The Main Theorem

The main theorem characterizing the FGC rings will be given. Then we will present a uniqueness of decomposition of finitely generated modules over FGC rings. Finally alternate forms and special cases of these results will be discussed.

Main Theorem 9.1: R is an FGC ring if and only if R is a finite product of rings of the following three types:

1. maximal valuation rings,
2. almost maximal Bezout domains, and
3. torch rings.

Proof: The three types of rings listed are FGC rings by 4.2, 5.2, and 5.4, and so a finite product of such rings is an FGC ring by 4.1(1) . •

Conversely, suppose R is an FGC ring. By 8.5 minspec$R$ is finite, say minspec$R = \{P_1,\dots,P_n\}$. If $M \in$ mspec$R$, then R_M is a valuation ring by 4.6 and 3.7, and so spec(R_M) is a chain. It follows that for $M \in$ mspec$R$, there is only one i such that $P_i \subset M$. In other words minspec$R$ is a set of pairwise comaximal ideals of R $(P_i + P_j = R$ for $i \neq j)$. Using the Chinese Remainder Theorem $R/\cap \text{spec}R = R/\bigcap_{i=1}^{n} P_i \cong \prod_{i=1}^{n} R/P_i$. $\cap$spec$R$ is the nilradical of R , and idempotents modulo the nilradical of R can be lifted to idempotents of R . Thus $R = \prod_{i=1}^{n} R_i$ where R_i is a ring with a unique minimal prime ideal. By 4.1(1) , each R_i is an FGC ring. It suffices to show that if R is an FGC ring with a unique minimal prime ideal, then R is one of the three types listed in the theorem.

Suppose R is an FGC ring with unique minimal prime ideal P . If R is local, then by 4.4 R is an almost maximal valuation ring. In this case if R is not a domain, then R is a maximal valuation ring by 3.3; and if R is a domain, then R is an almost maximal Bezout domain. Thus if R is a local ring, it is one of the three desired types of rings.

Suppose R is a non-local FGC ring with unique minimal prime ideal P . Consider the case where R is a domain, i.e., $P = \{0\}$. By 4.6 R is a locally almost maximal Bezout domain. Every non-zero prime ideal of R is a subset of a unique maximal ideal of R by 5.6. Let $x \in R - \{0\}$. R/Rx is an FGC ring by 4.1(2) , and so by 8.5 minspec(R/Rx) is finite. Thus there are only finitely many prime ideals of R minimal with respect to the property of containing x . Each maximal ideal of R which contains x , contains one of these minimal primes, and each non-zero minimal prime ideal containing x is a subset of only one maximal ideal of R . Therefore mspec (Rx) is finite . By definition, R is an h-local domain. R is a locally almost maximal h-local domain implies R is an almost maximal domain by 2.9 . Therefore R is an almost maximal Bezout domain, which is one of the three desired types of rings.

There remains to consider the case where R is a non-local FGC ring with a unique minimal prime ideal P and $P \neq \{0\}$. By 5.5 P is a uniserial R-module. R/P is an FGC ring by 4.1(2) , and so R/P is an h-local domain by the last paragraph. R is a locally almost maximal Bezout ring by 4.6. By definition, R is a torch ring, which is one of the three desired types of rings. q.e.d.

Theorem 9.2: (Uniqueness of decomposition) If R is an FGC ring and A is a finitely generated R-module, then the decomposition of A into a finite direct sum of indecomposable non-zero cyclic submodules is unique up to isomorphism. In other words, if $A = \bigoplus_{i=1}^{m} A_i = \bigoplus_{j=1}^{n} B_j$ where A_i and B_j are non-zero imdecomposable cyclic submodules of A , then $m = n$ and with a possible relabelling of the subscripts $A_i \cong B_i$ for all $i = 1,\ldots,m$.

Proof: By 9.1 R is a finite product of indecomposable FGC rings. The R-module A correspondingly decomposes into a direct sum, so it suffices to consider indecomposable FGC rings. Thus one must consider R being one of the three types of rings listed in 9.1 . In the first case, if R is a maximal valuation ring, then the uniqueness of decomposition follows from 3.4.

In the second case, one supposes R is an almost maximal Bezout domain. If $t(A)$ denotes the torsion submodule of A, then $A/t(A)$ is a finitely generated torsion-free R-module. By 3.9 $A/t(A)$ is a free R-module, say $A/t(A) \cong R^d$ for some $d \in N \cup \{0\}$. Then $A \cong t(A) \oplus A/t(A) \cong t(A) \oplus R^d$. If Q is the quotient field of R, then d is the dimension of $A \otimes_R Q$ as a Q-vector space. Thus d is independent of the decomposition of A, so there must be exactly d A_i's and d B_j's isomorphic to R. Thus we consider the case where A is a torsion R-module . R is h-local by 2.9, and so by 2.6 $A = \bigoplus_{M \in \mathrm{mspec} R} A(M)$. Using 2.5, R is h-local, and A is torsion it follows that each A_i is a submodule of $A(M)$ for some $M \in \mathrm{mspec}R$. By 2.7 $A(M)$ is an R_M-module, and each A_i is then an R_M-module for some $M \in \mathrm{mspec}R$. $A(M)$ is independent of the decomposition of A, so one is reduced to the case where $A = A(M)$ for some $M \in \mathrm{mspec}R$. $A(M)$ is an R_M-module and R_M is a valuation ring by 3.7, so the uniqueness of decomposition follows from 3.4.

In the third and last case, one supposes R is a torch ring. Let P be the unique minimal prime ideal of R, and using 5.3(7) there exists a unique $M \in \mathrm{mspec}R$ such that $P_M \cong P$ and $P_{M'} \cong \{0\}$ for all $M' \in \mathrm{mspec}R - \{M\}$. If I is an ideal of R, then by 2.1 $I = \bigcap_{M' \in \mathrm{mspec}R} I_{M'}{}^c$. If furthermore $I \subset P$, then $M' \in \mathrm{mspec}R - \{M\}$ implies $P_{M'} \cong \{0\}$, $I_{M'} = \{0\}$, and so $I_{M'}{}^c = P$. Thus $I = \bigcap_{M' \in \mathrm{mspec}R} I_{M'}{}^c$ and $I \subset P$ implies $I = I_M{}^c$. It follows that the correspondence $I \to I_M$ from $\{I: I \text{ is an ideal of } R \text{ and } I \subset P\}$ and $\{J: J \text{ is an ideal of } R_M \text{ and } J \subset P_M\}$ is one-to-one . Consider A_M, which is a finite direct sum of R_M-modules. R_M is a valuation ring by 3.7, so that decomposition of A_M

into a direct sum of cyclics is unique by 3.4 . By the above one-to-one correspondence the summands of A which are isomorphic to R/I for $I \subset P$ are independent of the decomposition. If I is an ideal of R with $I \not\subset P$, then $P \subsetneq I$ by 5.3(5) . In a decomposition of A into a direct sum of cyclics, the sum of the cyclic summands isomorphic to R/I with $I \not\subset P$ is isomorphic to the R/P-torsion submodule of A/PA . This is independent of the decomposition, and so these summands are unique since as an R/P-module this decomposition is unique by the earlier case where R is an almost maximal Bezout domain. This verifies the uniqueness of decomposition in the case R is a torch ring. <u>q.e.d.</u>

We next give an alternate form of the main theorem. This characterization of FGC rings is given in the paper by R. Wiegand and S. Wiegand [34], and has the advantage of describing FGC rings by one set of conditions.

<u>Theorem 9.3</u>: A ring is an FGC ring if and only if it is a finite product of rings R satisfying the following four conditions:

1. R has a unique minimal prime ideal P ,
2. P is a uniserial R-module,
3. R/P is an h-local Bezout domain, and
4. R is a locally almost maximal Prufer ring.

<u>Proof</u>: If R is an FGC ring, then by 9.1 R is a finite product of rings of three types, and it is straight forward to verify that each of these three types of rings satisfy the four conditions listed.

Conversely, suppose R satisfies the four conditions listed. If R is local, then R is an almost maximal valuation ring by (4) , and so R is an FGC ring by 4.2. If $P = \{0\}$, then R is an almost maximal Bezout domain by 2.9, and so R is an FGC ring by 5.2. We consider the remaining case where R is not local and $P \neq \{0\}$. Let $x \in R - P$ and $p \in P$. For $M \in \text{mspec}R$, one must have $R_M x \supset R_M p$ since R_M is a valuation ring. By 2.2(1) $Rx \supset Rp$.

It follows that R is a Bezout ring. Thus by definition R is a torch ring, and so R is an FGC ring by 5.4. This verifies that if R is a ring satisfying the four given conditions, then R is an FGC ring. If R is a finite product of rings satisfying the four conditions, then by 4.1(1) R is an FGC ring. q.e.d.

Restricting ones attention to domains, one gets the following special case of the main theorem. This characterization of FGC domains is a special case of the main theorem of the paper by W. Brandal and R. Wiegand [5] . The uniqueness of decomposition presented parallels the uniqueness of decomposition of finitely generated Abelian groups into a free Abelian group and a direct sum of p-primary groups.

Theorem 9.4: R is an FGC domain if and only if R is an almost maximal Bezout domain.

Suppose R is an FGC domain and A is a finitely generated R-module. Then there exist $m,n \in N \cup \{0\}$, distinct $\{M_1,\ldots,M_m\} \subset \text{mspec}R$, $\{k_1,\ldots,k_m\} \subset N$, and family of ideals of R $\{I_{ij}: j = 1,\ldots,m;\ i = 1,\ldots,k_j\}$ such that $\{0\} \neq I_{1j} \subset I_{2j} \subset \ldots \subset I_{k_j j} \subset M_j$, $\text{mspec}\ (I_{ij}) = \{M_j\}$, and $A \cong R^n \oplus (\bigoplus_{j=1}^{m} (\bigoplus_{i=1}^{k_j} R/I_{ij}))$. Moreover this decomposition is unique in the following sense: if also $A \cong R^{n'} \oplus (\bigoplus_{j=1}^{m'} (\bigoplus_{i=1}^{k'_j} R/I'_{ij}))$ where $\{0\} \neq I'_{1j} \subset I'_{2j} \subset \ldots \subset I'_{k_j j} \subset M'_j$ and $\text{mspec}\ (I'_{ij}) = \{M'_j\}$, then $m = m'$, $n = n'$, with a possible resubscripting of $\{M'_1,\ldots,M'_{m'}\}$, $M_j = M'_j$, $k_j = k'_j$, and $I_{ij} = I'_{ij}$ for all i,j .

Proof: By 9.1, R is an FGC domain if and only if R is an almost maximal Bezout domain.

Suppose R is an FGC domain and A is a finitely generated R-module. By repeating part of the proof of 9.2, one obtains $A \cong R^n \oplus (\bigoplus_{j=1}^{m} A(M_j))$ for some distinct $\{M_1,\ldots,M_m\} \subset \text{mspec}R$ and for each j, $A(M_j) \cong \overset{k_j}{\underset{i=1}{\oplus}} R/I_{ij}$ where $\text{mspec}(I_{ij}) = \{M_j\}$. If $1 \leq i$, $i' \leq k_j$ and $M \in \text{mspec}R - \{M_j\}$, then $(I_{ij})_M \cong R_M$; and so by 2.2(1) one deduces that $I_{ij} \subset I_{i'j}$ if and only if $(I_{ij})_{M_j} \subset (I_{i'j})_{M_j}$. R_{M_j} is a valuation domain, so it follows that I_{ij} and $I_{i'j}$ are comparable. With a possible resubscripting, one may assume $\{0\} \neq I_{1j} \subset I_{2j} \subset \ldots \subset I_{k_j j} \subset M_j$. This gives the existance of a decomposition of A.

R is an indecomposable R-module since R is a domain, and $\text{mspec}(I_{ij}) = \{M_j\}$ implies R/I_{ij} is an indecomposable R-module by 2.5. Thus all the indicated summands of A are indecomposable R-module. By 9.2 the decomposition is unique in the sense indicated. q.e.d.

We have commented that the Fundamental Theorem of Abelian Groups could be described as the statement that P.I.D.'s are FGC rings. If the Fundamental Theorem of Abelian Groups is derived by linear algebra techniques, one decomposes a finitely generated module over a P.I.D. R as follows: $A \cong R/Rd_1 \oplus \ldots \oplus R/Rd_n$ where the d_i's are non-unit elements of R with $d_2|d_1$, $d_3|d_2,\ldots, d_n|d_{n-1}$. The general ring theoretic version of this decomposition is called a canonical form decomposition.

Definition: Let A be an R-module. A canonical form decomposition of A is a decomposition $A \cong R/I_1 \oplus \ldots \oplus R/I_n$ where the ideals I_j of R satisfy $I_1 \subset I_2 \subset \ldots \subset I_n \neq R$.

We next give a decomposition theorem of FGC rings which is different than 9.2, namely that finitely generated modules over FGC rings have a unique canonical form decomposition. I. Kaplansky [14] studied canonical form decompositions, and showed that if they exist then they are unique.

Theorem 9.5: Let R be an FGC ring and A a finitely generated R-module. Then A has a canonical form decomposition, and such a decomposition is unique

Proof: We show that A has the decomposition by using 9.1 and considering the three types of indecomposable FGC rings.

Suppose R is a maximal valuation ring. By 9.2 A is a direct sum of indecomposable cyclics of the form R/I. Such a decomposition, with a possible reordering of the summands, is a canonical form decomposition since the ideals of R are totally ordered under set inclusion.

Suppose R is an almost maximal Bezout domain. By 9.4, $A \cong R^n \oplus (\bigoplus_{j=1}^{m} (\bigoplus_{i=1}^{k_j} R/I_{ij}))$ where $\{0\} \neq I_{1j} \subset I_{2j} \subset \ldots \subset I_{k_j j} \subset M_j$ and $\text{mspec}\,(I_{ij}) = \{M_j\}$. Let $r = \max\{k_j : j = 1,\ldots,m\}$ and for notational convenience set $I_{ij} = R$ for $k_j < i \leq r$. If I and J are ideals of R with $\text{mspec}\,(I) \cap \text{mspec}\,(J) = \emptyset$, then by the Chinese Remainder Theorem $R/I \oplus R/J \cong R/I \cap J$; and clearly $\text{mspec}\,(I \cap J) = \text{mspec}\,(I) \cup \text{mspec}\,(J)$. Thus $A \cong R/I_1 \oplus \ldots \oplus R/I_{n+r}$ where $I_1 = \ldots = I_n = \{0\}$, $I_{n+i} = \bigcap_{j=1}^{m} I_{ij}$ for $i = 1,\ldots,r$, and this is a canonical form decomposition of A.

Suppose R is a torch ring, with minimal prime ideal P. By 9.2 one can write $A = R/I_1 \oplus \ldots \oplus R/I_n \oplus T$ where $I_j \subset P$ for $j = 1,\ldots,n$ and T is a direct sum of cyclics of the form R/I where $I \not\subset P$. By 5.3(5) if $I \not\subset P$, then $P \subsetneq I$. Thus T is a finitely generated (torsion) R/P-module. R/P is an almost maximal Bezout domain, so T has a canonical form decomposition as an R/P-module by the last paragrpah. This is a canonical form decomposition of T as an R-module, and so A has a canonical form decomposition, since P is a uniserial R-module.

It has been verified that if R is an indecomposable FGC ring then A has a canonical form decomposition. If R is an FGC ring, then by 9.1 R is a finite product of indecomposable FGC rings, and A decomposes into a corresponding direct sum, each summand having a canonical form decomposition. In an obvious manner these can be combined to give a canonical form decomposition of A. This verifies that if R is an FGC ring, then a canonical form decomposition for A exists.

To prove uniqueness, suppose one has two canonical form decompositions $A \cong R/I_1 \oplus \ldots \oplus R/I_m \cong R/J_1 \oplus \ldots \oplus R/J_n$ where $I_1 \subset \ldots \subset I_m \neq R$ and $J_1 \subset \ldots \subset J_n \neq R$. For $M \in \text{mspec}R$, R_M is a valuation ring by 4.6 and 3.7. Choose $M \in \text{mspec}R$ such that $I_m \subset M$. Then A_M has m non-zero cyclic R_M-summands, and so by 3.4, the uniqueness of decomposition of A_M as an R_M-module, $m \leq n$. Similarly $n \leq m$ and so $m = n$. By 3.4 again, $(I_j)_M = (J_j)_M$ for all $j = 1,\ldots,m$ and for all $M \in \text{mspec}R$. By 2.1 $I_j = J_j$ for all $j = 1,\ldots,m$, and so the canonical form decomposition of A is unique. <u>q.e.d.</u>

Perhaps it has been noticed that the development has been one in which the integral domain case has been emphasized. The main theorem was proved by dealing with torch rings by reducing to the integral domain case, and the decomposition depended heavily on the properties of h-local domains. The paper by T. Shores and R. Wiegand [28] emphasized more the general ring case, and so the proofs there dealt more with canonical form decompositions; more like the treatment of I. Kaplansky [14].

R.S. Pierce [25] studied modules over (commutative) regular rings. It is not hard to show using the main theorem, 9.1, that the only indecomposable FGC regular rings are the fields. Thus as a special case of 9.1, one gets the theorem due to R.S. Pierce: If R is a (commutative) regular ring, then R is an FGC ring if and only if R is a finite product of fields.

PART 2 CONSTRUCTING EXAMPLES

Section 10 Valuations

G is a partially ordered group if G is an Abelian group with a partial ordering $\geq$ such that $x \geq y$ implies $x + z \geq y + z$ for all $x,y,z \in G$. G is a totally ordered group if G is a partially ordered group in which the partial ordering is a total ordering. If G is a partially ordered group, then we use G_+ to denote $\{g \in G: g \geq 0\}$. If G is an Abelian group and G_1 and G_2 are subsets of G, then $G_1 + G_2 = \{g_1 + g_2 \in G: g_1 \in G_1$ and $g_2 \in G_2\}$ and $-G_1 = \{-g_1 \in G: g_1 \in G_1\}$. If G and H are partially ordered groups and $f: G \to H$, then f is an order homomorphism if f is a group homomorphism and $x \leq y$ implies $f(x) \leq f(y)$ for all $x,y \in G$; and f is an order isomorphism if f is a group isomorphism and both f and f^{-1} are order homomorphisms. If X is a set with an additive group structure, then X^* will denote the set of all non-zero elements of X, i.e., $X^* = X - \{0\}$.

Lemma 10.1:

1. Let G be a totally ordered group. Then (i) $0 \in G_+$, (ii) $G_+ + G_+ \subset G_+$, and (iii) $\{G_+^*, \{0\}, -G_+^*\}$ is a partition of G.

2. Let G be an Abelian group with a subset P of G satisfying the conditions (i), (ii), and (iii) of part 1. If one defines $\geq$ on G by $x \geq y$ if $x - y \in P$ for $x,y \in G$, then $\geq$ makes G a totally ordered group.

3. If G is a totally ordered group, then G is a torsion-free Abelian group.

Proof: Straight forward.

Definition: Let K be a field and G a totally ordered group. Then the function $v: K^* \to G$ is a valuation if

(i) $v(xy) = v(x) + v(y)$ for all $x,y \in K^*$, and

(ii) $v(x + y) \geq \inf\{v(x), v(y)\}$ for all $x,y \in K^*$ such that $x + y \neq 0$.

If v is a valuation, we will follow the generally accepted convention that v must be a surjective function, i.e., $v(K^*) = G$.

Suppose $v: K^* \to G$ is a valuation. By definition v is a group homomorphism from the multiplicative group K^* to the additive group G , so $v(1) = 0$. Also $0 = v(1) = v((-1)^2) = 2v(-1)$. G is a torsion free group by 10.1(3) . Thus $v(-1) = 0$. If $x \in K^*$, then $v(-x) = v(-1) + v(x) = v(x)$ Thus if $x,y \in K^*$ with $x - y \neq 0$, then $v(x - y) \geq \inf\{v(x), v(y)\}$.

Suppose $v: K^* \to G$ is a valuation, $x,y \in K^*$, and $v(x) < v(y)$. Then $v(x + y) \geq \inf\{v(x), v(y)\} = v(x)$. On the other hand, $v(x) = v(x + y - y) \geq \inf\{v(x + y), v(y)\}$ and $v(x) < v(y)$ implies $v(x) \geq v(x + y)$. We have verified that $v(x) < v(y)$ implies $v(x + y) = v(x)$. In other words, if $v: K^* \to G$ is a valuation, $x,y \in K^*$, and $v(x) \neq v(y)$, then $v(x + y) = \inf\{v(x), v(y)\}$.

Let R be a domain with multiplicative group of units U and with Q the quotient field of R . Then Q^*/U is a multiplicative Abelian group. Define the ordering $\geq$ on Q^*/U by $aU \geq bU$ if $ab^{-1} \in R$ for $a,b \in Q^*$. Then Q^*/U is a partially ordered group, called the <u>divisibility group of R</u> . For example, if R is a valuation domain, then it is not difficult to see that R is a discrete rank one valuation domain if and only if the divisibility group of R is order isomorphic to Z .

<u>Lemma 10.2</u>:

1. If R is a valuation domain with quotient field Q , then the divisibility group of R is a totally ordered group and the canonical map of Q^* onto the divisibility group of R is a valuation.

2. Let $v: K^* \to G$ be a valuation. Define $R = \{0\} \cup \{x \in K^*: v(x) \geq 0\}$ Then R is a valuation domain, K is the quotient field of R , and G is order isomorphic to the divisibility group of R .

Proof: Straight forward.

This lemma amounts to the statement that valuations and valuation domains are just different ways of looking at the same things. If $v: K^* \to G$ is a valuation, then G is called the value group of v and the R of 10.2(2) is called the valuation domain of v. Of course this definition of valuation domain of v is consistent with our earlier definition of valuation ring. If R is a valuation domain and one uses the notation above, then the canonica map $Q^* \to Q^*/U$ is called the canonical valuation of R. Thus for a valuation domain R, the divisibility group of R is the value group of the canonical valuation of R.

If G is a totally ordered group, then I is an ideal of G if $I \subset G_+$ and ($x \in I$, $y \in G$, and $y \geq x$ implies $y \in I$). If G is a totally ordered group, then I is a prime ideal of G if I is an ideal of G, $I \neq G_+$, and $x,y \in G_+ - I$ implies $x + y \in G_+ - I$ (where the minus signs denote set complement). For example if G is a totally ordered group, then G_+^* and $\emptyset$ are always prime ideals of G.

Lemma 10.3: Let $v: K^* \to G$ be a valuation and R the valuation domain of v. Then there is a one-to-one correspondence between the set of ideals of R and the set of ideals of G given by $I \to v(I^*)$ for I an ideal of R. Under this correspondence prime ideals of R correspond to prime ideals of G.

Proof: Straight forward.

This last lemma can be generalized to Bezout domains. Namely if R is a Bezout domain, then there is a one-to-one correspondence between the ideals of R and the ideals of the divisibility group of R, and under this correspondence primes correspond to primes. The interested reader is refered to [4] for definitions and a more general discussion.

We will use <u>KdimR</u> to denote the <u>Krull dimension of R</u>, where of course this is one less than the length of the longest chain of prime ideals of R if such a longest finite chain exists, and infinity otherwise. Thus for a ring R, $\mathrm{Kdim}R \in \mathbb{N} \cup \{0,\infty\}$. If R is a valuation domain, then $\mathrm{Kdim}R = |\mathrm{spec}R| - 1$ if $|\mathrm{spec}R| < \infty$, and otherwise $\mathrm{Kdim}R = |\mathrm{spec}R| = \infty$. Similarly if G is a totally ordered group, we define the <u>Krull dimension of G</u>, denoted <u>KdimG</u>, by $\mathrm{Kdim}G = |\mathrm{spec}G| - 1$ if $|\mathrm{spec}G| < \infty$, and otherwise $\mathrm{Kdim}G = |\mathrm{spec}G| = \infty$; where of course <u>specG</u> denotes the set of prime ideals of G. If $v\colon K^* \to G$ is a valuation, then common terminology is to use "rank of v" to denote what we have called KdimG.

<u>Corollary 10.4</u>: If R is a valuation domain with divisibility group G, then $\mathrm{Kdim}R = \mathrm{Kdim}G$.

<u>Proof</u>: 10.2 and 10.3.

Another kind of dimension will also be needed. If G is an Abelian group and $g_1,\ldots,g_r \in G$, then $\{g_1,\ldots,g_r\}$ is said to be <u>rationally dependent</u> if there exist $n_1,\ldots,n_r \in \mathbb{Z}$ such that $\sum_{i=1}^{r} n_i g_i = 0$ and $n_i \neq 0$ for at least one i; and $\{g_1,\ldots,g_r\}$ is said to be <u>rationally independent</u> if it is not rationally dependent. If G is an Abelian group, then the <u>rational rank of G</u>, denoted <u>RrankG</u>, is the number of elements of a largest set of rationally independent elements of G if such a largest finite set exists, and otherwise $\mathrm{Rrank}G = \infty$. Thus if G is an Abelian group, then $\mathrm{Rrank}G \in \mathbb{N} \cup \{0,\infty\}$. Although we shall not need it in our development, it is not hard to see that the Krull dimension of a totally ordered group is less than or equal to its rational rank.

Let $G_1,\ldots,G_n$ be totally ordered groups and let $G = \prod_{i=1}^{n} G_i$. The <u>lexicographic ordering</u> on G is the ordering $\geq$ given by

$(x_1,\dots,x_n) \le (y_1,\dots,y_n)$ if $(x_1,\dots,x_n) = (y_1,\dots,y_n)$ or $x_i < y_i$ if $i = \inf\{j \in \{1,\dots,n\}: x_j \ne y_j\}$. The lexicographic ordering on G makes G a totally ordered group. In the case $n = 2$, then $(x_1,x_2) \le (y_1,y_2)$ if $x_1 < y_1$ or ($x_1 = y_1$ and $x_2 \le y_2$).

In our later discussion we shall consider the following example. Let $n \in N$ and for $1 \le i \le n$, let G_i be the additive group of rational numbers with the standard total ordering. Let $G = \prod_{i=1}^{n} G_i$ and let G have the lexicographic ordering. We claim that $\mathrm{Kdim}G = \mathrm{Rrank}G = n$. For define $P_i = \{(x_1,\dots,x_n) \in G_+^*: \inf\{j \in \{1,\dots,n\}: x_j > 0\} \le n+1-i\}$ for $i = 1,\dots,n$. Then $G_+^* = P_1 \supsetneqq P_2 \supsetneqq \dots \supsetneqq P_n \supsetneqq \emptyset$ and $\mathrm{spec}G = \{P_1,\dots,P_n, \emptyset\}$. This verifies that $\mathrm{Kdim}G = n$. Viewing G as an n-dimensional vector space over the field of rational numbers one easily concludes that $\mathrm{Rrank}G = n$.

Proposition 10.5: Let K be an algebraic field extension of k, let G' be a totally ordered group with subgroup G, and let $v': K^* \to G'$ and $v: k^* \to G$ be valuations such that $v'|k^* = v$. Then $\mathrm{Kdim}G' = \mathrm{Kdim}G$ and $\mathrm{Rrank}G' = \mathrm{Rrank}G$.

Proof: Let $g \in G'$. We claim that there exists $n \in N$ such that $ng \in G$. By convention $v'(K^*) = G'$, so there exists $x \in K^*$ such that $v'(x) = g$. x is algebraic over k, so $a_m x^m + a_{m-1}x^{m-1} + \dots + a_0 = 0$ for some $m \in N$, $a_i \in k$, and $a_m = 1$. If $v'(a_i x^i)$ are all different for $i = 0,1,\dots,m$ then $v'(a_m x^m + \dots + a_0) = \inf\{v'(a_i x^i): i = 0,\dots,m\}$ and so $a_m x^m + \dots + a_0$ cannot be zero, Thus there exist i and j, $0 \le i < j \le m$, with $v'(a_i x^i) = v'(a_j x^j)$, and of course $a_i \ne 0$, $a_j \ne 0$. Then $(j-i)\,v'(x) = v'(a_i) - v'(a_j) = v(a_i) - v(a_j) \in G$. If $n = j - i$, then $n \in N$ and $ng \in G$ as claimed.

To show that the Krull dimensions are the same, define $f: \mathrm{spec}G' \to \mathrm{spec}G$ by $f(P') = P' \cap G$ for $P' \in \mathrm{spec}G'$, and define $h: \mathrm{spec}G \to \mathrm{spec}G'$ by

$h(P) = \{g' \in G'$: there exists $n \in N$ such that $ng' \in P\}$ for $P \in \mathrm{spec}G$. It is straight forward using the first paragraph to show that f and h are functions with the appropriate ranges and that f and h are inverses of each other. Hence $\mathrm{Kdim}G' = \mathrm{Kdim}G$.

Clearly $\mathrm{Rrank}G' \geq \mathrm{Rrank}G$. Let $\{g_1,\ldots,g_r\}$ be a rationally independent subset of G'. By the first paragraph, there exist $n_1,\ldots,n_r \in N$ such that $n_i g_i \in G$. Then $\{n_1g_1,\ldots,n_rg_r\}$ is a rationally independent subset of G. Thus $\mathrm{Rrank}G' \leq \mathrm{Rrank}G$, and so $\mathrm{Rrank}G' = \mathrm{Rrank}G$. q.e.d.

Proposition 10.6: Let $v: k^* \to G$ be a valuation and let K be a field extension of k. Then there exists a valuation $v': K^* \to G'$ such that $v'|k^* = v$, where of course $G' \supset G$.

Proof: Let $v: k^* \to G$ be the given valuation, let R be the valuation domains of v, and let M be the maximal ideal of R. Let $\mathcal{S} =$ $\{S$: S is a ring, $R \subset S \subset K$, and $MS \neq S\}$. $R \in \mathcal{S}$, so $\mathcal{S} \neq \emptyset$. Order $\mathcal{S}$ by set inclusion. By Zorn's Lemma $\mathcal{S}$ has a maximal element, say V is a maximal element of $\mathcal{S}$.

Let $x \in K^*$. We wish to show that $x \in V$ or $1/x \in V$. Suppose $x \notin V$ and $1/x \notin V$. Then $V \neq V[x]$ and since V is a maximal element of $\mathcal{S}$, one must have $MV[x] = V[x]$. Similarly $MV[1/x] = V[1/x]$. Then one can write $1 = \sum_{i=0}^{m} a_i x^i$ and $1 = \sum_{j=0}^{n} b_j/x^j$ for $a_i, b_j \in MV$. We suppose that these relations are of the smallest length m and n. Without loss of generality we may assume $m \geq n$. Multiply the first equation by $1 - b_0$ and the second equation by $a_m x^m$ getting

$$1 - b_0 = (1 - b_0)a_0 + \ldots + (1 - b_0)a_m x^m \quad \text{and}$$

$$(1 - b_0)a_m x^m = a_m b_1 x^{m-1} + \ldots + a_m b_m x^{m-n} .$$

Combining,

$$1 - b_0 = (1 - b_0)a_0 + \ldots + (1 - b_0)a_{m-1}x^{m-1} + a_m b_1 x^{m-1} + \ldots + a_m b_m x^{m-n} ,$$

i.e., $1 = \sum_{i=0}^{m-n} c_i x^i$ where $c_i \in MV$ and $m - n < m$. This contradicts the minimality of m . Thus we have verfied the claim that if $x \in K^*$, then $x \in V$ or $1/x \in V$. By 3.1 V is a valuation domain, and K is the quotient field of V . Let P be the maximal ideal of V . $V \in \mathcal{S}$ implies $MV \neq V$, $MV \subset P$ and in particular $M \subset P$. Hence $P \cap k = M$ and $V \cap k = R$. Let G' be the divisibility group of V and let v' be the canonical valuation of V . Using $P \cap k = M$, $V \cap k = R$, and an appropriate identification, one can view v' as the required extension of v . <u>q.e.d.</u>

<u>Definition</u>: If R is a local ring with maximal ideal M , then the field R/M is called the <u>residue field</u> of R .

<u>Theorem 10.7</u>: (W. Krull [17]) Let k be a field and G a totally ordered group. Then there exists a valuation domain R such that the residue field of R is isomorphic to k and the divisibility group of R is order isomorphic to G .

<u>Proof</u>: Let S be the group algebra $k[G]$. S is a domain. Let Q be the quotient field of S . Define $v\colon Q^* \to G$ by

$$v\left(\sum_{i=1}^{m} c_i X_{g_i} / \sum_{j=1}^{n} c_j' X_{g_j'}\right) = \inf\{g_i : i = 1,\ldots,m\} - \inf\{g_j' : j = 1,\ldots,n\} ,$$ where

it is assumed that $c_i \in k^*$ for all i, $g_i \neq g_j$ if $i \neq j$, and similarly for the expression in the denominator. v is a valuation. Let R be the valuation domain of v . Q is the quotient field of R , k is isomorphic to the residue field of R , and G is order isomorphic to the divisibility group of R . <u>q.e.d.</u>

Alternate proofs of 10.7 use $S = k[G_+]$ as in page 107 of [1], or

$S = k[X_g : g \in G]$, where X_g are indeterminants over k , as in 18.5 of [9] .

The next few facts are about valuation domains. The first fact is that a finite intersection of valuation domains gives a Bezout domain. To prove this one needs the following technical lemma. The proofs of 10.8 and 10.9 are taken from N. Bourbaki [1] .

Lemma 10.8: Let Q be a field and for $i = 1,\dots,n$ let V_i be a valuation domain with maximal ideal P_i , and Q is the quotient field of V_i . Let $x \in Q^*$. Then there exists $f(X) \in Z[X]$ with $f(X)$ of the form $f(X) = X^k + n_{k-1}X^{k-1} + \dots + n_1X + 1$, $k \geq 2$, such that (i) $f(x) \neq 0$, (ii) if $x \in V_i$, then $f(x)$ is a unit of V_i, and (iii) if $x \notin V_i$, then $xf(x)^{-1} \in P_i$.

Proof: Let $I = \{i \in \{1,\dots,n\}: x \in V_i\}$. For $i \in I$, let x_i denote the image of x in the residue field V_i/P_i . If there exists $g(X) \in Z[X]$ with leading and last coefficient 1 (i.e., $g(X)$ is of the same form as the desired $f(X)$) such that $g(x_i) = 0$, then choose $f_i(X)$ to be this $g(X)$. If no such $g(X)$ exists, then choose $f_i(X)$ to be 1 . Define $f(X) = 1 + X^2 \prod_{i \in I} f_i(X)$. Then $f(x) \neq 0$. If $x \in V_i$, $i \in I$, $f(x_i) \neq 0$, and so $f(x)$ is a unit of V_i . Finally suppose $x \notin V_i$, $i \notin I$. Then

$$V_if(x)^{-1} = V_i[1/x^k(1 + \frac{n_{k-1}}{x} + \frac{n_{k-2}}{x^2} + \dots + \frac{1}{x^k})] = V_i[1/x^k] \text{ since } 1/x \in P_i .$$

Thus $V_ixf(x)^{-1} = V_ix(1/x^k) = V_i(1/x^{k-1}) \subset P_i$ since $k \geq 2$, and so $xf(x)^{-1} \in P_i$. q.e.d.

Proposition 10.9: Let Q be a field and for $i = 1,\dots,n$ let V_i be a valuation domain with maximal ideal P_i , $V_i \subset Q$, and Q is the quotient field of V_i . Suppose $V_i \not\subset V_j$ if $i \neq j$. Define $R = \bigcap_{i=1}^{n} V_i$ and $M_i = R \cap P_i$. Then R is a Bezout domain, $\text{mspec}R = \{M_1,\dots,M_n\}$, $M_i \neq M_j$

if $i \neq j$, $R_{M_i} = V_i$, and Q is the quotient field of R .

Proof: Consider $i = 1$. It is clear that $M_1 \in \operatorname{spec}R$. We verify that $R_{M_1} = V_1$. Clearly $R_{M_1} \subset V_1$. Let $x \in V_1^*$. Apply 10.8 to find an appropriate $f(X) \in Z[X]$. Then for all $i = 1,\ldots,n$ $f(x)^{-1} \in V_i$ and $xf(x)^{-1} \in V_i$. Hence $f(x)^{-1}$, $xf(x)^{-1} \in R$. $x \in V_1^*$, so $f(x)^{-1}$ is a unit of V_1 . Hence $x = xf(x)^{-1}/f(x)^{-1} \in R_{M_1}$. This verifies $R_{M_1} = V_1$, and similarly for $i = 2,\ldots,n$. If $i \neq j$ and $M_i \subset M_j$, then $V_i = R_{M_i} \supset R_{M_j} = V_j$, contrary to the assumption. This verifies that the M_i's are distinct. Q is the quotient field of V_1 and $R_{M_1} = V_1$, so Q is the quotient field of R . If $u \in R - \bigcup_{i=1}^{n} M_i$, then u is a unit of V_i for all $i = 1,\ldots,n$ and so $u^{-1} \in \bigcap_{i=1}^{n} V_i = R$. Hence $R - \bigcup_{i=1}^{n} M_i$ is the set of units of R , and therefore $\operatorname{mspec}R = \{M_1,\ldots,M_n\}$. By definition R is a semilocal Prufer domain, so by 3.8 R is a Bezout domain. q.e.d.

Proposition 10.10: Let R be a valuation domain with Q the quotient field of R . Then:

1. R is a maximal domain if and only if Q is a linearly compact R-module.
2. If R is a maximal domain and $P \in \operatorname{spec}R$, then R_P is a maximal valuation domain.

Proof:

1. If Q is a linearly compact R-module, then R being an R-submodule of Q is also a linearly compact R-module (1.2(1)) , and hence by definition R is a maximal domain. Conversely suppose R is a maximal domain, i.e., R is a linearly compact R-module. Let $\{x_\alpha + A_\alpha\}_{\alpha \in X}$ be a family of cosets of submodules of Q ($x_\alpha \in Q$ and A_α is an R-submodule of Q) with the

f.i.p. . We may assume $A_\beta \neq Q$ for some $\beta \in X$, for otherwise $\bigcap_{\alpha \in X} x_\alpha + A_\alpha = Q \neq \emptyset$, as desired. For this β , choose $b \in Q$ such that $b \notin A_\beta$ and $Rb \supset Rx_\beta$. Then $\frac{x_\beta}{b} + \frac{1}{b} A_\beta \subset R$. $\{(\frac{x_\alpha}{b} + \frac{1}{b} A_\alpha) \cap (\frac{x_\beta}{b} + \frac{1}{b} A_\beta)\}_{\alpha \in X}$ is a family of cosets of R-submodules of R with the f.i.p., and since R is a linearly compact R-module, $\emptyset \neq \bigcap_{\alpha \in X} ((\frac{x_\alpha}{b} + \frac{1}{b} A_\alpha) \cap (\frac{x_\beta}{b} + \frac{1}{b} A_\beta)) = \bigcap_{\alpha \in X} \frac{x_\alpha}{b} + \frac{1}{b} A_\alpha$. It follows that $\emptyset \neq \bigcap_{\alpha \in X} x_\alpha + A_\alpha$, and so Q is a linearly compact R-module.

2. Suppose R is a maximal ring and $P \in \operatorname{spec} R$. Then Q is the quotient field of R_p . Q is a linearly compact R-module by part 1 , so clearly Q is a linearly compact R_p-module. By part 1 again, R_p is a maximal ring. Clearly R_p is a valuation domain. q.e.d.

10.10(2) is more generally true for valuation rings which are not domains. For a proof see the paper by D.T. Gill [7] .

Section 11 Long Power Series Rings

We want to define long power series rings. These are discussed in O. Schilling's text [26]. The name "long power series ring" was suggested by L. Levy.

Let K be a field and G a totally ordered group. K^G will denote the set of all functions from G to K. If $f \in K^G$, then define the <u>support of f</u>, denoted <u>$\text{sppt}(f)$</u>, by $\text{sppt}(f) = \{x \in G: f(x) \neq 0\}$. Define $Q = \{f \in K^G: \text{sppt}(f) \text{ is a well ordered subset of } G\}$ and $R = \{f \in Q: \text{sppt}(f) \subset G_+\}$. R is called a <u>long power series ring</u>, or more specifically the <u>long power series ring relative to K and G</u>, and Q is the associated overring.

To motivate the terminology, we describe an alternate method of considering long power series rings. In our proofs we shall need to consider both descriptions. With K and G as above, consider the set of all <u>long power series</u> of the form $\sum_{\alpha \in W} c_{g_\alpha} X_{g_\alpha}$ where W is the set of all ordinals less than some fixed ordinal (the fixed ordinal may be different for different long power series), $g_\alpha \in G$, $\alpha < \beta$ implies $g_\alpha < g_\beta$, and $c_{g_\alpha} \in K$. The X_g for $g \in G$ are considered as indeterminants. The standard convention about polynomials is used, i.e., two long power series are considered equal if they differ only in terms of the form $0X_g$ where $0 \in K$ and $g \in G$. Then one can identify Q with the set of all long power series as follows: if $f \in Q^*$, then there is an order isomorphism $\phi: W \to \text{sppt}(f)$ where W is the set of all ordinals less than some ordinal, and so f is identified with the long power series $\sum_{\alpha \in W} f(\phi(\alpha)) X_{\phi(\alpha)}$.

To define the operations of addition and multiplication on long power series rings, we shall need the following two lemmas.

Lemma 11.1: Let G be a totally ordered group and $h \in G$. Let S and T be non-empty well ordered subsets of G. Then $U = \{s \in S:$ there exists $t \in T$ such that $s + t = h\}$ is a finite set.

Proof: Let $s_1 = \inf U$, and recursively $s_{n+1} = \inf(U - \{s_1,\ldots,s_n\})$. If at some finite stage $U - \{s_1,\ldots,s_n\} = \emptyset$, then the proof is complete. Thus suppose this process of defining the s_n's does not stop after a finite number of steps. For each $n \in N$ choose $t_n \in T$ such that $s_n + t_n = h$. $s_1 < s_2 < \ldots$, and so $t_1 > t_2 > \ldots$. But it is impossible to have an infinite decreasing sequence in a well order set. Contradiction. q.e.d.

Lemma 11.2: Let G be a totally ordered group and $H \subset G$. Let S and T be non-empty well ordered subsets of G. Define $U = \{s + t \in G: s \in S$, $t \in T$, and $s + t > h$ for all $h \in H\}$, and assume $U \neq \emptyset$. Then $\inf U$ exists and $\inf U \in U$.

Proof: Suppose this result is not true. Then there exists $\{u_n\}_{n\in N} \subset U$ such that $u_n > u_{n+1}$ for all $n \in N$. For each $n \in N$ choose $s_n \in S$ and $t_n \in T$ such that $s_n + t_n = u_n$. Using the well orderedness of S, the sequence $\{a(n)\}_{n\in N} \subset N$ is defined inductively as follows: $a(1)$ is chosen so that $s_{a(1)} = \inf\{s_i\}_{i\in N}$, and having chosen $a(n)$, then $a(n+1)$ is chosen so that $s_{a(n+1)} = \inf\{s_{a(n)+i}\}_{i\in N}$. Then $a(1) < a(2) < \ldots$, $s_{a(1)} < s_{a(2)} < \ldots$, and $t_{a(1)} > t_{a(2)} > \ldots$. But it is impossible to have an infinite decreasing sequence in a well ordered set. Contradiction. q.e.d.

Proposition 11.3: Let K be a field and G a totally ordered group. Let R be the long power series ring relative to K and G, and let Q be the associated overring. For $f,g \in Q$, define the sum $f + g$ and the product fg by $(f + g)(x) = f(x) + g(x)$ for all $x \in G$, and $(fg)(x) = \sum\{f(y)g(z): y,z \in G$ and $y + z = x\}$ for all $x \in G$. With these operations Q is a domain and R is a subring.

Proof: It is clearly the case that $f + g$ makes sense, $\mathrm{sppt}(f + g)$ is a well ordered subset of G, and so $f + g \in Q$.

In the definition of the product fg it is to be understood that only non-zero terms in the sum contribute to the sum. For the sum to be meaningful, it must contain only a finite number of non-zero terms. By 11.1 with $h = x$, $S = \text{sppt}(f)$, and $T = \text{sppt}(g)$ one deduces that the sum has only a finite number of non-zero terms. Thus $(fg)(x)$ is a well defined element of K for all $x \in G$, and so $fg \in K^G$. We need to verify that $\text{sppt}(fg)$ is a well ordered subset of G. Suppose X is a non-empty subset of $\text{sppt}(fg)$. Let $H = \{h \in G: h < x \text{ for all } x \in X\}$. Apply 11.2 with $S = \text{sppt}(f)$ and $T = \text{sppt}(g)$. Then it is the case that $X \subset U$, $\inf U$ exists and $\inf U \in U$. If $\inf U \notin X$, then $\inf U \in H$ and this contradicts the definition of U, namely $u \in U$ implies $u > h$ for all $h \in H$. Thus $\inf U \in X$ and so $\inf U = \inf X$ is the required least element of X. This verifies that $\text{sppt}(fg)$ is a well ordered subset of G and hence $fg \in Q$.

It is a straight forward but tedious task to verify that with these operations Q is a domain and R is a subring of Q. <u>q.e.d.</u>

In the next two proofs it will be convenient to view the elements of long power series rings as long power series instead of as functions. In each case a particular element must be constructed using transfinite induction.

<u>Proposition 11.4</u>: Let K be a field and G a totally ordered group. Let R be the long power series ring relative to K and G with associated over-ring Q. Then:

1. If $f \in R$ and $0 \in \text{sppt}(f)$, then f is a unit of R.
2. R is a valuation domain and Q is the field of fractions of R.
3. The divisibility group of R is order isomorphic to G and the residue field of R is isomorphic to K.

<u>Proof</u>:

1. Let $f \in R$ and suppose $0 \in \text{sppt}(f)$. We identify elements of R

with long power series, and so $f = \sum_{\alpha \in W} c_{g_\alpha} X_{g_\alpha}$. $0 \in \text{sppt}(f)$ means that one may assume that $g_1 = 0$ and $c_{g_1} \neq 0$. We must construct a $g = \sum_{\alpha \in W'} d_{h_\alpha} X_{h_\alpha} \in R$ such that $fg = 1$. The h_α and d_{h_α} are defined by transfinite induction. Let $h_1 = 0$ and $d_{h_1} = (c_{g_1})^{-1}$. Suppose α is an ordinal, $\alpha > 1$, and h_β and d_{h_β} have been defined for all $\beta < \alpha$. Then $f \cdot \sum_{\beta < \alpha} d_{h_\beta} X_{h_\beta}$ is an element of R with 1 as the constant term (really $1X_0$ where $1 \in K$ and $0 \in G$) so the product can be written in the form $1 + \sum_{\gamma \in W''} e_{i_\gamma} X_{i_\gamma}$. If $\{i_\gamma : \gamma \in W''$ and $e_{i_\gamma} \neq 0\} = \emptyset$ then we have our desired inverse of f. Otherwise define $h_\alpha = \inf\{i_\gamma : \gamma \in W'' \text{ and } e_{i_\gamma} \neq 0\}$ and $d_{h_\alpha} = -(c_{g_i})^{-1} e_{h_\alpha}$. Then $f \cdot \sum_{\beta \leq \alpha} d_{h_\beta} X_{h_\beta} = 1 + \sum_{\gamma \in W''} e_{i_\gamma} X_{i_\gamma} + f \cdot (-(c_{g_i})^{-1} e_{h_\alpha} X_{h_\alpha})$ and this last expression is 1 plus a sum of terms with the smallest X_g term appearing (with a non-zero coefficient) being strictly larger than X_{h_α}. In other words, the h_α's so defined have the property that $h_\alpha > h_\beta$ if $\alpha > \beta$. This process of defining the h_α and d_{h_α} must eventually stop and the $g = \sum_{\alpha \in W'} d_{h_\alpha} X_{h_\alpha}$ so defined for an appropriate W' is an element of R and is the desired inverse of f.

2. We shall return to the use of functions to represent elements of Q and R. For $x \in G$, define $f_x \in Q^*$ by $f_x(y) = 1$ if $x = y$ and $f_x(y) = 0$ if $x \neq y$, for $y \in G$. In other words, $f_x(y) = \delta_{xy}$, the Kronecker delta function. In this notation, the multiplicative identity of R and Q is f_0 where $0 \in G$.

If $f \in Q^*$, let $x = \inf(\text{sppt}(f))$. Then one can write $f = f_x g$ for $g \in R$ and $0 \in \text{sppt}(g)$. By part (1), this g is a unit of R. Considering

Q as an R-module if $x,y \in G$, then $f_x | f_y$ if and only if $y \geq x$. It follows that R is a valuation domain and Q is the quotient field of R.

3. Let U denote the set of units of R. Then $U = \{f \in R: 0 \in \text{sppt}(f)\}$ by part (1). The divisibility group of R is by definition Q^*/U with the ordering $aU \geq bU$ if $ab^{-1} \in R$. Define $i: Q^*/U \to G$ by $i(qU) = \inf(\text{sppt}(q))$ for $q \in Q^*$. Then i is well defined and i is an order isomorphism. This verifies that the divisibility group of R is order isomorphic to G. Define $j: R \to K$ by $j(f) = f(0)$ for $f \in R$. Then j is a ring epimorphism and $\ker j$ is the maximal ideal M of R. Hence R/M, the residue field of R, is isomorphic to K. <u>q.e.d.</u>

Let R be the long power series ring relative to K and G, and let Q be the quotient field of R. If one identifies the divisibility group of R with G as in the above proof, then the canonical valuation of R is $v: Q^* \to G$ where $v(f) = \inf(\text{sppt}(f))$ for $f \in Q^*$.

<u>Proposition 11.5</u>: Every long power series ring is a maximal valuation domain.

<u>Proof</u>: By 11.4 every long power series ring is a valuation domain. Let R be the long power series ring relative to K and G. Elements of R will be considered as long power series instead of as functions. We must show that R is a linearly compact R-module. Let $\{x_i + I_i\}_{i \in W}$ be a family of cosets of submodules of R with the f.i.p., and it must be shown that $\bigcap_{i \in W} x_i + I_i \neq \emptyset$. We will construct by transfinite induction $g_\alpha \in G$ and $c_{g_\alpha} \in K$ for all ordinals α less than some fixed ordinal, so that $f = \sum_\alpha c_{g_\alpha} X_{g_\alpha} \in \bigcap_{i \in W} x_i + I_i$. If $x_i \in I_i$ for all $i \in W$, then $f = 0$ is the desired element in the intersection. Thus chose $i \in W$ such that $x_i \notin I_i$. Then write $x_i = \sum_{\beta \in V} d_{h_\beta} X_{h_\beta}$ and one may assume $d_{h_1} \neq 0$. Define $g_1 = h_1$ and $c_{g_1} = d_{g_1}$. Suppose α is an ordinal, $\alpha > 1$, and g_β and c_{g_β} have been defined for all ordinals $\beta < \alpha$.

If $\sum_{\beta<\alpha} c_{g_\beta} X_{g_\beta} \in \bigcap_{i\in W} x_i + I_i$, then we have found our desired f . Otherwise there exists $i \in W$ such that $\sum_{\beta<\alpha} c_{g_\beta} X_{g_\beta} \notin x_i + I_i$. Write $\sum_{\beta<\alpha} c_{g_\beta} X_{g_\beta} - x_i = \sum_{\gamma\in V'} e_{h'_\gamma} X_{h'_\gamma}$ and one may assume that $e_{h'_1} \neq 0$. Define $g_\alpha = h'_1$ and $c_{g_\alpha} = e_{h'_1}$. By the f.i.p. of $\{x_i + I_i\}_{i\in W}$ it follows that $g_\beta < g_\alpha$ if $\beta < \alpha$. Hence this process of defining g_α and c_{g_α} must eventually stop, and one gets the desired element f in the intersection. q.e.d.

We shall not have a need to consider the topology on valuation domains induced by the valuation. But those readers familiar with this valuation topology will probably have noticed that the elements constructed in the last two proofs are limits of appropriate Cauchy sequences. In other words, a basic property being used is that long power series rings are complete in the valuation topology. To be more complete in this topic, by 11.5 long power series rings are maximal valuation domains. And for R a valuation domain, R is a maximal ring if and only if R is an almost maximal ring and R is complete in the valuation topology. For a proof of this last statement, see the paper by E. Matlis [21] .

Section 12 Maximally Complete Valuation Domains

The use of long power series rings will be needed to construct some of the examples of FGC rings. One fact needed is 12.8, that for certain K and G, the long power series ring relative to K and G has the property that its quotient field is an algebraically closed field. For this we will need to consider the concept of a maximally complete valuation domain, which will be motivated by giving some examples.

<u>Example 12.1</u>: Let k be a field and let G be the additive group of rational numbers with the usual total ordering. Use the Krull construction of the proof of 10.7 to construct the corresponding valuation domain. That is, let S be the group algebra $k[G]$ and let Q_1 be the quotient field of S. Define $v_1 : S^* \to G$ by $v_1(\sum_{i=1}^{n} c_i X_{g_i}) = \inf\{g_i : c_i \neq 0\}$. Extend v_1 to Q_1^*, also denoted v_1, by $v_1(s/s_1) = v_1(s) - v_1(s_1)$ for $s, s_1 \in S^*$. v_1 is a valuation. Let R_1 be the valuation domain of v_1, i.e., $R_1 = \{0\} \cup \{x \in Q_1^* : v_1(x) \geq 0\}$. By 10.7 the divisibility group of R_1 is order isomorphic to G and the residue field of R_1 is isomorphic to k.

<u>Example 12.2</u>: This example appears in the O. Zariski and P. Samuel text [37] on page 101. Let k and G be as in the last example. Let

$$Q_2 = \{\sum_{i \in N} c_i X_{g_i} : c_i \in k ,\ g_i \in G ,\ \text{if } i < j \text{ then } g_i < g_j ,\ \text{and } \lim_{i \to \infty} g_i = \infty\}.$$

Of course one uses the standard convention that two series (elements of Q_2) are equal if they differ by most terms of the form $0X_g$ for $0 \in k$ and $g \in G$. With the standard formal operations Q_2 becomes a field. For example $(X_1 + X_2)^{-1} = \sum_{i \in N} (-1)^{i+1} X_{i-2}$. Define $v_2 : Q_2^* \to G$ by

$v_2(\sum_{i \in N} c_i X_{g_i}) = \inf\{g_i : c_i \neq 0\}$. Then v_2 is a valuation. Let R_2 be the valuation ring of v_2, i.e., $R_2 = \{0\} \cup \{x \in Q_2^* : v_2(x) \geq 0\}$. By 10.2(2), the divisibility group of R_2 is order isomorphic to G and the residue field of R_2 is isomorphic to k .

Using the notation of the last two examples, Q_1 can be embedded into Q_2 in an obvious manner. With identification, we have $v_2|Q_1{}^* = v_1$, and so R_1 is then embedded into R_2 . We claim that in this embedding, the image of R_1 is not all of R_2 . For example, $X_1/(X_1 + X_2) \in R_1$ and the image of this element in R_2 is $\sum_{i \in N} (-1)^{i+1} X_{i-1}$, so that "infinite series" may appear as the image of elements of R_1 . However if $\sum_{i \in N} c_i X_{g_i}$ is the image of an element of R_1 , and $c_i \neq 0$ for all $i \in N$, then by studying the process of computing inverse in Q_2 , we see that $\{g_i\}_{i \in N}$ is a sequence of rational numbers that after a finite number of terms increases "linearly" Thus $\sum_{i \in N} X_{i!}$ is an element of R_2 which is not an image of an element of R_1 .

Example 12.3: Let k and G be as in the last two examples. Let R_3 be the long power series ring relative to k and G , and let Q_3 be its quotient field . By 11.4(3) the divisibility group of R_3 is order isomorphic to G and the residue field of R_3 is isomorphic to k .

Using the notation of the last three examples and viewing elements of Q_3 as long power series, Q_2 can be embedded into Q_3 in an obvious manner. Let v_3 be the canonical valuation of R_3 , then with an identification $v_3|Q_2{}^* = v_2$, and so R_2 is then embedded into R_3 . The image of R_3 of R_2 is not all of R_3 since $\sum_{i \in N} X_{1-1/i}$ is an element of R_3 which is not in the image of R_2 .

One can summarize these three examples by the following commutative diagram

$$\begin{array}{ccccc} R_1 & \longrightarrow & R_2 & \longrightarrow & R_3 \\ \downarrow & & \downarrow & & \downarrow \\ Q_1 & \longrightarrow & Q_2 & \longrightarrow & Q_3 \end{array}$$

of rings, where all the maps are proper embeddings. R_1, R_2, and R_3 are all valuation domains with isomorphic divisibility groups and isomorphic residue fields, and Q_1, Q_2, and Q_3 are their respective quotient fields.

We wish to indicate a sense in which R_3 and Q_3 form the last column in the above diagram, namely how R_3 is a "maximally complete" valuation domain.

Example 12.4: Let k and G be as in the last three examples. Let K be a proper field extension of k. Let R_4 be the long power series ring relative to K and G, and let Q_4 be the field of fractions of R_4. Then the divisibility group of R_4 is order isomorphic to G and the residue field of R_4 is isomorphic to K by 11.4(3). We obtain a commutative diagram of rings, where all the maps are proper embeddings

$$\begin{array}{ccc} R_3 & \longrightarrow & R_4 \\ \downarrow & & \downarrow \\ Q_3 & \longrightarrow & Q_4 \end{array} .$$

Example 12.5: Let k and G be as in the last four examples. Let G' be a totally ordered group properly containing G. For example G' could be the additive group of real numbers or G' could be $G \oplus G$ ordered lexicographically and G is identified as a subgroup of G' via the embedding $g \to (g,0)$ for $g \in G$. Let R_5 be the long power series ring relative to k and G' , and let Q_5 be the field of fractions of R_5 . Then the divisibility group of R_5 is order isomorphic to G' and the residue field of R_5 is isomorphic to k by 11.4(3) . We get a commutative diagram of rings where all the maps are proper embeddings

$$\begin{array}{ccc} R_3 & \longrightarrow & R_5 \\ \downarrow & & \downarrow \\ Q_3 & \longrightarrow & Q_5 \end{array} .$$

Going from R_1 to R_2 and going from R_2 to R_3 we maintained the same divisibility group and residue field, but going from R_3 to R_4 the residue field enlarged, and going from R_3 to R_5 the divisibility group enlarged.

Definition: Let $R \to R'$ be an embedding of the valuation domain R into the valuation domain R' . Then $R \to R'$ gives R' as an immediate extension of R if the divisibility groups and the residue fields of R and R' are isomorphic via this embedding.

Generally one says that R' is an immediate extension of R , when there is only one embedding being considered. In the above examples, R_2 is an immediate extension of R_1 and R_3 is an immediate extension of R_2 . Both R_4 and R_5 are not immediate extensions of R_3 .

Definition: Let R be a valuation domain. Then R is a maximally complete valuation domain if R has no proper immediate extensions.

According to I. Kaplansky [13], this definition is due to F.K. Schmidt but first published by W. Krull in 1932. The next result shows that for valuation domains maximally complete is the same as maximal. The proof is taken from the text by O. Schilling [26] .

Theorem 12.6: Let R be a valuation domain. Then R is a maximally complete valuation domain if and only if R is a maximal domain.

Proof: Suppose R is a maximal domain and R is not a maximally complete valuation domain. Let M be the maximal ideal of R, G the divisibility group of R, Q the quotient field of R, and $v: Q^* \to G$ the canonical valuation. Since R is not a maximally complete valuation domain there exists a proper immediate extension R' of R with corresponding M', G', Q', and v' as for R above. To simplify notation, it is assumed that $R \to R'$ and $Q \to Q'$ are inclusion maps, $G = G'$ and $v'|Q^* = v$. The extension is proper means there exists an $s \in R' - R$. Let $X = \{r \in R^*: v(r) \geq v'(s)\}$ and let $Y = \{v'(s - r): r \in X\}$.

We wish to show that Y does not have a largest element. Let $r \in X$. We show that there exists $q \in R^*$ such that $r + q \in X$ and $v'(s - (r + q)) > v'(s - r)$. There exists $c \in R^*$ such that $v'(s-r) = v(c)$, and so $(s - r)c^{-1}$ is a unit of R'. The inclusion map $R \to R'$ induces an isomorphism of $R/M \to R'/M'$, so there exists a $d \in R - M$ such that $(s - r)c^{-1} + M' = d + M'$, and so $v'((s - r)c^{-1} - d) > 0$. $v'(s - r - cd) = v'((s - r)c^{-1} - d) + v'(c) > v'(c) = v'(s - r)$. So $q = cd$ has the property that $v'(s - r - q) > v'(s - r)$. $v(r + q) \geq v(q) = v'(q) = v'(s - r) \geq v'(s) \geq 0$, implying $q \in R^*$ and $r + q \in X$. We have shown that Y does not have a largest element.

For $r \in X$, let $I_r = \{0\} \cup \{t \in R^*: v(t) \geq v'(s - r)\}$. Consider the family of congruences $\{x \equiv r \bmod I_r\}_{r \in X}$. Every finite subfamily of these congruence is solvable, so the entire family is solvable since R is a maximal valuation domain. Let r_0 be a solution, i.e., $r_0 \in R$ and $r_0 \equiv r \bmod I_r$

for all $r \in X$. If $r \in X$ and $r \neq r_0$, then $v(r_0 - r) \geq v'(s - r)$ and so $v'(s - r_0) = v'(s - r + r - r_0) \geq \inf\{v'(s - r), v'(r - r_0)\} = v'(s - r) \geq \inf\{v'(s), v'(r)\} = v'(s)$. Hence $r_0 \in X$ and $v'(s - r_0) \geq v'(s - r)$ for all $r \in X$. So $v'(s - r_0)$ is the largest element of Y , contradicting the last paragraph. We have proved half of theorem 12.6 .

Conversely, suppose R is not a maximal valuation domain. Let M be the maximal ideal of R , G the divisibility group of R , Q the quotient field of R , and $v: Q^* \to G$ the canonical valuation. There exists $x_\alpha \in R$ and ideals I_α of R such that $\bigcap_{\alpha \in X} x_\alpha + I_\alpha = \emptyset$ yet $\{x_\alpha + I_\alpha\}_{\alpha \in X}$ has the f.i.p..

Well order the index set X , and by possibly throwing away some of the cosets $x_\alpha + I_\alpha$ we may assume $x_\alpha + I_\alpha \subset x_\beta + I_\beta$ if $\alpha \geq \beta$ and $I_\alpha \subset M$ for all $\alpha \in X$. Let z be an indeterminant over Q . Define $S = \{f(z) \in Q[z]^*:$ there exists $\gamma(f) \in X$ such that $\alpha \in X$ and $\alpha \geq \gamma(f)$ implies $v(f(x_\alpha)) = v(f(x_{\gamma(f)}))$, and in particular $f(x_\alpha) \neq 0\}$.

Case 1: Assume $S = Q[z]^*$. Define $w: Q[z]^* \to G$ by $w(f(z)) = v(f(x_{\gamma(f)}))$ for $f(z) \in Q[z]^*$. For $f(z), g(z) \in Q[z]^*$, $w(f(z)g(z)) = w(f(z)) + w(g(z))$, and if also $f(z) + g(z) \neq 0$, then $w(f(z) + g(z)) \geq \inf\{w(f(z)), w(g(z))\}$. Extend w to $Q(z)^*$, also denoted w , by $w: Q(z)^* \to G$ with $w(f(z)/g(z)) = w(f(z)) - w(g(z))$ for $f(z)$, $g(z) \in Q[z]^*$. Then w is a valuation and identifying Q as the constants of $Q(z)$ $w|Q^* = v$. Let R_w be the valuation domain of w and M_w the maximal ideal of R_w . We wish to show that R_w is a proper immediate extension of R . Clearly it is a proper extension, and the divisibility groups $w(Q(z)^*)$ and $v(Q^*)$ are the same. We shall show that the inclusion map $R \to R_w$ induces an isomorphism, $R/M \to R_w/M_w$, of the residue fields. For this it suffices to show that $R + M_w \supset R_w$. We first show that if $f(z) \in Q[z] \cap (R_w - M_w)$, then there exists $r_f \in R - M$ and $m_f \in M_w$ such that $f(z) = r_f + m_f$. Let $f(z) \in Q[z] \cap (R_w - M_w)$, and let

$\alpha > \gamma(f)$. Define $b = x_\alpha - x_{\gamma(f)}$. $x_\alpha \equiv x_{\gamma(f)} \bmod I_{\gamma(f)}$ and $I_{\gamma(f)} \subset M$ implies $b \in M$. Expanding polynomials one gets $v(f(x_\alpha) - f(x_{\gamma(f)}))$ $= v(f(x_{\gamma(f)} + b) - f(x_{\gamma(f)})) = v(\sum_{i=1}^{n} f_i(x_{\gamma(f)})\, b^i) \geq v(b) > 0$ for some $f_i(z) \in Q[z]$. By the definition of w , $w(f(z) - f(x_{\gamma(f)})) > 0$. Let $r_f = f(x_{\gamma(f)})$ and $m_f = f(z) - f(x_{\gamma(f)})$, and so $f(z) = r_f + m_f$. To show $R + M_w \supset R_w$, let $f(z)/g(z) \in R_w - M_w$. By possibly multiplying and dividing by an element of Q^* , we may assume $f(z), g(z) \in Q[z] \cap (R_w - M_w)$. By the earlier statement, there exist $r_f, r_g \in R - M$ and $m_f, m_g \in M_w$ such that $f(z) = r_f + m_f$ and $g(z) = r_g + m_g$. Then $f(z)/g(z) = (r_f + m_f)/(r_g + m_g) =$ $(r_f/r_g) + (r_g m_f - r_f m_g)/(r_g r_g + r_g m_g)$ where $r_f/r_g \in R - M$ and $(r_g m_f - r_f m_g)/(r_g r_g - r_g m_g) \in M_w$. We have shown that $R/M \to R_w/M_w$ is an isomorphism , and so R_w is a proper immediate extension of R . Therefore R is not a maximally complete valuation domain.

Case 2: Assume $S \neq Q[z]^*$. Let $f(z) \in Q[z]^* - S$ and assume $f(x_\gamma) \notin I_\gamma$ for some $\gamma \in X$. Let $\alpha \in X$ with $\alpha > \gamma$, and let $b = x_\alpha - x_\gamma$ $x_\alpha \equiv x_\gamma \bmod I_\gamma$ implies $b \in I_\gamma$. Expanding polynomials, $v(f(x_\alpha) - f(x_\gamma)) =$ $v(\sum_{i=1}^{n} f_i\,(x_\gamma)\, b^i) \geq v(b) \in v(I_\gamma^*)$. Thus $f(x_\alpha) - f(x_\gamma) \in I_\gamma$, and since $f(x_\gamma) \notin I_\gamma$ it follows that $v(f(x_\alpha)) = v(f(x_\gamma))$. Thus $f(z) \in S$, contrary to the assumption. We have shown that if $f(z) \in Q[z]^* - S$, then $f(x_\gamma) \in I_\gamma$ for all $\gamma \in X$. Choose $f_0(z) \in Q[z]^* - S$ with the property that if $f(z) \in Q[z]^* - S$, then $\deg f(z) \geq \deg f_0(z)$. Since $f_0(x_\alpha) \in I_\alpha$ for all $\alpha \in X$ and $\bigcap_{\alpha \in X} x_\alpha + I_\alpha = \emptyset$, we deduce that $\deg f_0(z) \geq 2$. We claim that $f_0(z)$ is irreducible in $Q[z]$. For suppose not, with $f_0(z) = f_1(z)\, f_2(z)$, $f_1(z), f_2(z) \in Q[z]$, $\deg f_1(z) \geq 1$, and $\deg f_2(z) \geq 1$. By the choice of $f_0(z)$ having the lowest degree, $f_1(z), f_2(z) \in S$. Let

$\gamma_0 = \max\{\gamma(f_1), \gamma(f_2)\}$. If $\alpha \geq \gamma_0$, then $v(f_0(x_\alpha)) = v(f_1(x_\alpha)) + v(f_2(x_\alpha)) = v(f_1(x_{\gamma_0})) + v(f_2(x_{\gamma_0})) = v(f_0(x_{\gamma_0}))$, and so $f_0(z) \in S$, a contradiction. This verifies that $f_0(z)$ is irreducible in $Q[z]$. Let u be an element of some algebraic field extension of Q with u a root of $f_0(z) = 0$. Consider the field extension $Q[u] \cong Q(u) \cong Q[z]/(f_0(z))$. An arbitrary element of $Q[u]$ can be written as $g(u)$ where $\deg g(z) < \deg f_0(z)$, and so $g(z) \in S$. Define $w: Q[u]^* \to G$ by $w(g(u)) = v(g(x_{\gamma(g)}))$. w is a valuation of $Q[u]$ extending v , and repeating part of the argument given in the first case, we see that the valuation ring of w is a proper immediate extension of R . Therefore R is not a maximally complete valuation domain. q.e.d.

Corollary 12.7: If R is a long power series ring, then R is a maximally complete valuation domain.

Proof: 11.5 and 12.6 .

Refering to the examples 12.1 through 12.5 , R_3 is a long power series ring and hence is a maximally complete valuation domain by 12.7 . Thus R_3 and Q_3 form the last possible column in the earlier commutative diagram in the sense that any added column consisting of proper embeddings must have the property that the residue field or the divisibility group is larger than that for R_3 .

Theorem 12.8: Let K be an algebraically closed field. Let $n \in N$ and let G be the direct sum of n copies of the additive group of rational numbers with the standard total ordering, and where G has the lexicographic ordering. Let R be the long power series ring relative to K and G , with Q the quotient field of R . Then Q is an algebraically closed field.

Proof: Let $\overline{Q}$ be an algebraic closure of Q . By 11.4 and 10.2(1) there is a valuation $v: Q^* \to G$ with R the valuation domain of v . By 10.6 there exists a totally ordered group $G' \supset G$, with compatible ordering, and a valuation $\overline{v}: \overline{Q}^* \to G'$ such that $\overline{v}|Q^* = v$. Let $\overline{R}$ be the valuation domain of $\bar{v}$. By 10.5 $\operatorname{Rrank} G = \operatorname{Rrank} G'$. G is a divisible Abelian group and so G is a direct summand of G' . G' is a torsion-free Abelian group since every totally ordered group is torsion-free by 10.1(3) . Hence if $G \neq G'$, then $\operatorname{Rrank} G' > \operatorname{Rrank} G$, contradicting an earlier statement. It follows that $G = G'$, and so the divisibility group of $\bar{R}$ is order isomorphic to the divisibility group of R . Since $\bar{Q}$ is an algebraic field extension of Q , the residue field of $\bar{R}$ is an algebraic field extension of the residue field of R . But the residue field of R is isomorphic to K by 11.4(3) , and K is an algebraically closed field. Thus the residue field of $\bar{R}$ is isomorphic to the residue field of R . Hence the inclusion $R \to \bar{R}$ makes $\bar{R}$ an immediate extension of R . R is maximally complete by 12.7, and so $R = \bar{R}$. Thus $Q = \bar{Q}$ and Q is an algebraically closed field. q.e.d.

As an additional comment for 12.8, if $K = C$, the field of complex numbers, and $n = 1$, then $Q \cong C$, as will be shown in the proof of 14.4 .

This section is closed with a few additional comments about immediate extensions and maximally complete valuation domains. Using Zorn's Lemma to get a "maximal" immediate extension, it follows that if R is a valuation domain, then there exists an immediate extension R' of R such that R' is a maximally complete valuation domain. See [26] for details of a proof - the main difficulty is getting an upper bound on the cardinality of an immediate extension of R . I. Kaplansky [13] showed that if R is a valuation domain then this "maximal" immediate extension of R is unique up to isomorphism if the characteristic of the residue field of R is zero, and by an example this "maximal" immediate extension of R need not be unique up to isomorphism if the characteristic of the residue field of R is not zero. Also studied is the question of when a maximally complete valuation domain is a long power series ring.

Section 13 Examples of Maximal Valuation Rings

The main theorem 9.1 states that the FGC rings are exactly the rings which are finite direct products of maximal valuation rings, almost maximal Bezout domains, and torch rings. The next three sections present examples of the indecomposable FGC rings of these three types.

We begin by considering the maximal valuation domains. Fields are maximal valuation domains. Besides fields, perhaps the best known examples of maximal valuation domains are the p-adic integers, for p a prime integer. The divisibility group of the p-adic integers is order isomorphic to Z with the standard total ordering, and the residue field of the p-adic integers is isomorphic to the field Z/pZ . If R is the long power series ring relative to Z/pZ and Z , then it is not hard to see that R is isomorphic to the p-adic integers.

More generally long power series rings are maximal valuation domains by 11.5; and given any totally ordered group and any field, there is a long power series ring with that divisibility group and that residue field by 11.4(3) . For examples of maximal (i.e., maximally complete) valuation domains other than long power series rings, the reader is refered to the paper by I. Kaplansky [13] or the text by O. Schilling [26] .

Examples of maximal valuation rings which are not domains, include R/I for R a maximal valuation domain and I an ideal of R , I not a prime ideal of R . For example Z/p^nZ for p a prime integer and $n \in N - \{1\}$. Other examples include quotients of long power series rings (of which Z/p^nZ is a special case).

Section 14 Examples of Almost Maximal Bezout Domains

It was noticed in section five that P.I.D.'s are almost maximal Bezout domains. Included in the list of P.I.D.'s are fields, Z, Z_{pZ} for p a prime integer, the p-adic integers, and polynomial rings in one variable over a field. Other almost maximal Bezout domains include the maximal valuation domains of the last section.

In 1952 the only known FGC domains were the P.I.D.'s and the almost maximal valuation domains. In fact I. Kaplansky in his text [16], page 80, remarked as to whether these were the only FGC domains. In 1973 W. Brandal [2], and independently in 1974 T. Shores and R. Wiegand [28], gave an examples of an FGC domain which is not a P.I.D. and not a valuation domain. This example will be presented in 14.1 . This domain first appeared in a paper by E. Matlis [24] and is credited to B. Osofsky. E. Matlis was studying domains with the property that every torsion-free module of finite rank decomposes into a direct sum of modules of rank one. Such domains are called domains with property D, and this example is a domain with property D.

Example 14.1: (B. Osofsky [24]) There exists an FGC domain R which is neither a P.I.D. nor a valuation domain. This R is an almost maximal Bezout domain with exactly two maximal ideals M_1 and M_2 such that R_{M_1} and R_{M_2} are both maximal valuation domains of Krull dimension one.

Proof: Let C be the field of complex numbers and let G be the additive group of rational numbers with the standard total ordering. Let R_1 be the long power series ring relative to C and G, with Q the quotient field of R_1. If one views elements of Q as long power series, then in an obvious manner one can consider the polynomial ring $C[X]$ as a subring of R_1, and hence $C(X)$ as a subfield of Q. Define $\alpha: C(X) \to C(X)$ by $\alpha(f(X)/g(X)) = f(X+1)/g(X+1)$ for $f(X), g(X) \in C(X)$, $g(X) \neq 0$. Then α is a field automorphism of $C(X)$. Let Y be a transcendence basis of Q over $C(X)$. Then the identity map of Y yields a field automorphism $\beta: C(X)(Y) \to C(X)(Y)$ extending α . Q is an algebraic extension of $C(X)(Y)$ and Q is an algebraically closed field by 12.8 , so there exists a field automorphism $\gamma: Q \to Q$ extending β . Clearly $\gamma|C = 1_C$ where C is viewed as the constants of $C(X)$, and hence is a subfield of Q . Also $\gamma(X) = X + 1$. Let $R_2 = \gamma(R_1)$ and let $R = R_1 \cap R_2$. By 11.4(1) $X - 1$ is a unit of R_1 and so $1/(X-1) \in R_1$. $\gamma(1/(X-1)) = 1/X \notin R_1$. Hence $R_2 = \gamma(R_1) \not\subset R_1$. Similarly $R_1 \not\subset R_2$. Since $R_1 \cong R_2$ are valuation domains, by 10.9 R is a Bezout domain with exactly two maximal ideals M_1 and M_2 such that $R_{M_1} \cong R_1$ and $R_{M_2} \cong R_2 \cong R_1$, and Q is the quotient field of R . Since $\mathrm{Kdim}R_1 = 1$, it must be the case that $\mathrm{spec}R = \{M_1, M_2, \{0\}\}$. Hence R is h-local by definition. R is a locally maximal valuation domain, so by 2.9 R is an almost maximal domain. Thus R is an FGC domain by 5.2 or the main theorem 9.1 . R is not a valuation domain since R has two maximal ideals. R is not a P.I.D. since the maximal ideal of $R_1 \cong R_{M_1}$ is not a finitely generated ideal. <u>q.e.d.</u>

A generalization of example 14.1 appears in the paper by T. Shores and R. Wiegand [28] . Namely if $n \in N - \{1\}$, then there exists an almost maximal Bezout domain with exactly n maximal ideals and every localization at a maximal ideal is a maximal valuation domain of Krull dimension one. The construction is similar to the one for 14.1 except one uses n "independent" automorphisms of Q . The details of this construction will not be given here since it is a special case of the example due to S. Wiegand which is presented next.

A <u>chain</u> of a partially ordered set X is a totally ordered subset of X . A <u>tree</u> is a partially ordered set X such that $\{y \in X: y \leq x\}$ is a chain of X for all $x \in X$. A <u>zero</u> of a partially ordered set X is an element

z of X such that $z \leq x$ for all $x \in X$. The zero of a partially ordered set, if it exists, will be denoted 0. If R is a Prufer ring, then specR is a tree with respect to the partial ordering of set inclusion. If R is a Prufer domain (or more specifically if R is a Bezout domain), then specR is a tree with zero.

Example 14.2: (S. Weigand [35]) If X is a finite tree with zero, then there exists a Bezout domain R such that specR is order isomorphic to X and R_P is a maximal valuation domain for all $P \in$ specR.

Proof: Let $\{C_1, C_2, \ldots, C_n\}$ be the set of distinct maximal chains of X. For $i = 1, \ldots, n$ let $C_i = \{c_{i0}, c_{i1}, \ldots, c_{ik_i}\}$ where $0 = c_{i0} < c_{i1} < \ldots < c_{ik_i}$. We may by a possible relabelling assume $k_1 \geq k_2 \geq \ldots \geq k_n$. Let $k = k_1$. Note that it is possible that $c_{ia} = c_{jb}$ for $i \neq j$, but in this case one must have $a = b$ and then $c_{ir} = c_{jr}$ for $r = 0, 1, \ldots, a$. For each $i = 2, 3, \ldots, n$ choose $r_i \in \{1, 2, \ldots, i-1\}$ such that $|C_i \cap C_{r_i}| \geq |C_i \cap C_j|$ for all $j = 1, 2, \ldots, i-1$. Let $s_i = |C_i \cap C_{r_i}|$. Then $s_i \in \{1, 2, \ldots, k_i\}$, $c_{ij} = c_{r_i j}$ for all $j = 0, 1, \ldots, s_i - 1$ and $c_{is_i} \neq c_{r_i s_i}$. Then the set of C_i, c_{ij}, r_i, and s_i completely determine the tree X. In particular $X = \bigcup_{i=1}^{n} C_i = \bigcup_{i=1}^{n} (\bigcup_{j=0}^{k_i} \{c_{ij}\})$.

Let G_0 be the additive group of rational numbers with the standard total ordering. Let $G = G_0^k$, where G has the lexicographic ordering. Elements of G will be written as $(g_1, \ldots, g_k)$ for $g_i \in G_0$. Let C be the field of complex numbers. Let R_1 be the long power series ring relative to C and G, and let Q be the quotient field of R_1.

We view the elements of Q as functions, i.e., elements of C^G. As done earlier, if $x \in G$, then $f_x \in C^G$ is defined by $f_x(y) = \delta_{xy}$ for $y \in G$. f_0 is the multiplicative identity of Q, where $0 \in G$. We define

$y_1 = f_{(1,0,\dots,0)}$, $y_2 = f_{(0,1,0,\dots,0)}, \dots,$ $y_k = f_{(0,0,\dots,0,1)}$. Let $C' = \{cf_0 : c \in C\}$. Then C' is a subring of R_1 which is isomorphic to the field C . Let S be the subfield of Q given by $S = C'(y_1,\dots,y_k)$. In the obvious manner S is isomorphic to $C(X_1,\dots,X_k)$ where $X_1,\dots,X_k$ are indeterminants over C . Thus we have fields $C' \subsetneq S \subsetneq Q$.

Let $v: Q^* \to G$ be given by $v(f) = \inf(\mathrm{sppt}(f))$ for $f \in Q^*$. Then the valuation domain of v is R_1 . For $j = 1,\dots,k$ let $P_j = \{0\} \cup \{x \in R_1^*:$ $v(x)$ has a positive entry in the first j coordinates$\}$. Let $P_0 = \{0\} \subset R_1$. Then $\mathrm{spec}R_1 = \{P_j : j = 0,1,\dots,k\}$ and $P_0 \subsetneq P_1 \subsetneq \dots \subsetneq P_k$. Also $y_j \in P_j - P_{j-1}$ for $j = 1,\dots,k$ and $1 = f_0 \in R_1 - P_k$.

We wish to inductively define field automorphisms $\alpha_1,\dots,\alpha_n$ of Q . α_1 is the identity automorphism of Q . Let $i \in \{2,3,\dots,n\}$ and suppose $\alpha_1,\dots,\alpha_{i-1}$ have been defined. Let α_i' be the field automorphism of S satisfying $\alpha_i'|C' = 1_{C'}$ and $\alpha_i'(y_j) = \alpha_{r_i}(y_j)$ for $j = 1,2,\dots,s_i-1$ and $\alpha_i'(y_j) = y_j + if_0$ for $j = s_i,\dots,k$. Let Y be a transcendence basis of Q over S . Let α_i'' be the field automorphism of $S(Y)$ such that $\alpha_i''|S = \alpha_i'$ and $\alpha_i''|Y = 1_Y$. Q is an algebraic extension of $S(Y)$ and Q is an algebraically closed field by 12.8 . Thus there exists a field automorphism α_i of Q such that $\alpha_i|S(Y) = \alpha_i''$. This completes the construction of the α_i's .

For $i = 1,\dots,n$ let $R_i = \alpha_i((R_1)_{P_{k_i}})$. Define $R = \bigcap_{i=1}^{n} R_i$. We wish to verify that this R has the required properties. Clearly Q is the quotient field of R_i and R_i is a valuation domain for all $i = 1,\dots,n$. Let M_i be the maximal ideal of R_i , i.e., $M_i = \alpha_i(P_{k_i}(R_1)_{P_{k_i}})$.

Consider $i \in \{2,3,\dots,n\}$ and let $x_i = \alpha_i(y_{k_i}) = y_{k_i} + if_0$. For $j \neq i$,

$0 \in \text{sppt}(\alpha_j^{-1}(x_i))$ and 11.4(1) implies x_i is a unit of R_j. $x_i \in M_i$. Hence $x_i^{-1} \in (R_1 \cap \ldots \cap R_{i-1} \cap R_{i+1} \cap \ldots \cap R_n) - R_i$. Similarly $y_k \in (R_2 \cap \ldots \cap R_n) - R_1$. Hence $R_i \not\subset R_j$ if $i \neq j$.

Let $M_i' = R \cap M_i$. By 10.9 R is a Bezout domain, mspec$R = \{M_1',\ldots,M_n'\}$, $M_i' \neq M_j'$ if $i \neq j$, $R_{M_i'} = R_i$ for all $i = 1,\ldots,n$, and Q is the quotient field of R. If $P \in \text{spec}R$, then there exists $i \in \{1,\ldots,n\}$ and $j \in \{0,1,\ldots,k_i\}$ such that $R_P = (R_i)_{P_j}$. R_1 is a maximal ring by 11.5, R_i is a maximal ring by 10.10(2), and so R_P is a maximal ring by 10.10(2) again. This verifies that if $P \in \text{spec}R$, then R_P is a maximal valuation domain.

It suffices to show that X is order isomorphic to spec$R$. Define $\phi: X \to \text{spec}R$ by $\phi(c_{ir}) = R \cap \alpha_i(P_r(R_1)_{P_{k_i}})$ for $c_{ir} \in X$. It is tedious but straight forward to verify that the construction has been such that ϕ is a well defined order isomorphism of X onto spec$R$. <u>q.e.d.</u>

It should be noted that all the domains constructed in 14.2 are not FGC domains. Namely, if X is a finite tree with zero and there exists $x \in X - \{0\}$ and $y,z \in X$ such that $y > x$, $z > x$, and y and z are not comparable, then the constructed Bezout domain R of 14.2 is not h-local, and by 9.1 and 2.9 this R is not an FGC domain. Another way to describe this condition of X is to say that there exist two distinct maximal chains of X whose intersection properly contains $\{0\}$.

<u>Example 14.3</u>: (S. Wiegand [35]) Suppose X is a finite tree with zero such that the intersection of any two distinct maximal chains of X is $\{0\}$. Then there exists an FGC domain R such that X is order isomorphic to spec$R$. Moreover R_P is a maximal valuation domain for all $P \in \text{spec}R$.

<u>Proof</u>: Apply the construction of 14.2 to get a Bezout domain R such that X is order isomorphic to spec$R$ and R_P is a maximal valuation domain for

all $P \in \text{specR}$. Hence R is a locally almost maximal ring. By the hypothesis on X and X is order isomorphic to specR it must be the case that R is h-local. By 2.9 R is an almost maximal domain, and so by 5.2 or the main theorem 9.1, R is an FGC domain. q.e.d.

Of course example 14.1 is a special case of example 14.3. The next example is a generalization of the S. Wiegand construction in the sense that an FGC domain R exists with infinitely many maximal ideals such that every localization is a maximal valuation domain.

Example 14.4: (P. Vamos [30]) There exists an FGC domain R with countably infinite many maximal ideals such that R_M is a Krull dimension one maximal valuation domain for all $M \in \text{mspecR}$.

Proof: Let C be the field of complex numbers, and let P be the prime subfield of C , i.e., P is the field of rational numbers. Let G be the additive group of rational numbers with the standard total ordering. Let Q be the quotient field of the long power series ring relative to C and G . Recall that we use c for the cardinality of the set of all real numbers.

Let T be a transcendence basis of Q over P . $c = c^{|N|} = |C|^{|N|} = |C^N| \leq |Q| \leq |C^G| = c$, and so $|Q| = c$. Q is an algebraic field extension of the field $P(T)$, so $|P(T)| = |Q| = c$ and hence $|T| = c$. Q is an algebraically closed field by 12.8 . Hence Q and C are both algebraically closed fields with c being the cardinality of a transcendence basis over P . Therefore $Q \cong C$ as fields.

Let $\{T_n\}_{n\in N}$ be a partition of T such that $|T| = |T_n|$ for all $n \in N$. For $n \in N$ let Q_n be the algebraic closure of $P(\bigcup_{i=1}^{n} T_i)$ in Q . $Q_n \subsetneq Q_{n+1}$ for all $n \in N$ and $Q = \bigcup_{n\in N} Q_n$. As in the last paragraph $Q_n \cong C$ for all

$n \in N$. Let Q_n' be the quotient field of the long power series ring relative to Q_n and G . As in the last paragraph there exists a field isomosphism $f: Q_n' \to Q$ such that $f|Q_n = 1_{Q_n}$. This isomorphism f and canonical valuation of $Q_n'^*$ onto G induces a valuation $v_n: Q^* \to G$. Let R_n be the valuation domain of v_n . Then R_n is a maximal valuation domain by 11.5, Q is the quotient field of R_n , $v_n(Q_n^*) = \{0\}$, and in particular $Q_n \subset R_n$. Moreover, we can assume that f is chosen such that $v_n(Q_{n+1}^*) \neq \{0\}$. Let $R = \bigcap_{n\in N} R_n$. We claim that this is the desired R .

If $x \in Q^*$, then $x \in Q_n^*$ for some n . Since $Q_i \subset Q_{i+1}$ and $v_i(Q_i^*) = \{0\}$, it follows that $v_i(x) = 0$ for all but finitely many $n \in N$, and we describe this condition by the statement that $\{v_n\}_{n\in N}$ is a family of valuations of Q with finite character.

We next wish to show that the intersection $R = \bigcap_{n\in N} R_n$ is irredundant in the sense that $R_n \not\subset \bigcap_{i\in N-\{n\}} R_i$ for all $n \in N$. Let $n \in N$. Since $v_n(Q_{n+1}^*) \neq \{0\}$, choose $x \in Q_{n+1}^* - Q_n$ such that $v_n(x) \neq 0$. By possibly replacing x by $1/x$, we may assume that $v_n(x) < 0$. Then $x \in (\bigcap_{i\in N-\{n\}} R_i) - R_n$, showing that the intersection is irredundant.

We verify that Q is the quotient field of R . Let $x \in Q$. Then $x \in Q_n$ for some $n \in N - \{1\}$. Q_n is the quotient field of $R_i \cap Q_n$ for $i = 1,\ldots,n-1$. As in the last paragraph $R_i \cap Q_n \not\subset R_j \cap Q_n$ for $i \neq j$ and $i,j = 1,\ldots,n-1$. By 10.9 Q_n is the quotient field of the domain $\bigcap_{i=1}^{n-1} (R_i \cap Q_n)$. Since $R_i \supset Q_n$ for $i \geq n$, $\bigcap_{i=1}^{n-1} (R_i \cap Q_n) = R \cap Q_n$. Thus $x = y/z$ for $y,z \in R \cap Q_n$, $z \neq 0$. In particular x is in the quotient field of R . This verifies that Q is the quotient field of R .

We next show that R is a Bezout domain. Let $x_1,\ldots,x_k \in R$. Then there exists $n \in N - \{1\}$ such that $x_1,\ldots,x_k \in Q_n$. As in the last paragraph $x_1,\ldots,x_k \in R \cap Q_n = \bigcap_{i=1}^{n-1} (R_i \cap Q_n)$ and by 10.9 $R \cap Q_n$ is a Bezout domain. It follows that $Rx_1 + \ldots + Rx_k$ is a principal ideal of R and so R is a Bezout domain.

Let P_n be the maximal ideal of R_n and let $M_n = P_n \cap R$. We wish to show that $\text{mspec}R = \{M_n\}_{n\in N}$. Let $n \in N$. Clearly M_n is a proper ideal of R. Suppose I is an ideal of R and $I \supsetneq M_n$. There exists $x \in (\bigcap_{i\in N-\{n\}} R_i) - R_n$ since the intersection $R = \bigcap_{n\in N} R_n$ is irredundant. R is a Bezout domain and Q is the quotient field of R implies there exists $x' \in Q$ such that $Rx' = Rx + R$. Let $p = 1/x'$. Then $p \in M_n$ and p is a unit of R_i for all $i \in N - \{n\}$. Let $y \in I - M_n$. R is a Bezout domain implies there exist $u \in I$ such that $Ru = Rp + Ry$. u is a unit of R_i for all $i \in N$, and so u is a unit of R. Thus $R = Ru = Rp + Ry \subset I$ and so $I = R$. It follows that $\text{mspec}R \supset \{M_n\}_{n\in N}$. On the other hand suppose that $M \in \text{mspec}R$, and we wish to show that $M = M_n$ for some $n \in N$. Suppose this is not the case, i.e., $M \neq M_n$ for all $n \in N$. There exists $x_1 \in M - M_1$. $\{v_n\}_{n\in N}$ is a family of valuations of Q with finite character implies x_1 is a unit of R_n for all $n > k$ for some $k \in N$. There exist $x_i \in M - M_i$ for $i = 2,3,\ldots,k$. R is a Bezout domain implies there exists $g \in M$ such that $Rg = Rx_1 + \ldots + Rx_k$. Then $g \notin M_n$ for all $n \in N$ and so g is a unit of R_n for all $n \in N$, and hence g is a unit of R. Then $M = R$ and this contradicts $M \in \text{mspec}R$. This verifies $\text{mspec}R = \{M_n\}_{n\in N}$, and clearly the M_n's are distinct.

We next verify that $R_{M_n} = R_n$ for all $n \in N$. $R_{M_n} = R_{R\cap P_n} \subset (R_n)_{P_n} = R_n$, so $R_{M_n} \subset R_n$. On the other hand suppose $x \in R_n$. R is a Bezout domain implies there exists $u \in Q$ such that $Ru = Rx + R$. $1 \in R \subset R$ implies $1/u \in R$. $0 = v_n(1) = v_n(u) + v_n(1/u)$, $v_n(u) \geq 0$, implies $v_n(1/u) = 0$. Thus $1/u \in R - M_n$. $x \in R - M_n$. $x \in Rx \subset Rx + R = Ru$ implies $x = ru$ for some $r \in R$. Thus $x = ru = r/(1/u)$ where $r \in R$ and $1/u \in R - M_n$. Thus $x \in R_{M_n}$, $R_n \subset R_{M_n}$, and so $R_{M_n} = R_n$. It follows that if $M \in$ MspecR, then $M = M_n$ for some $n \in N$, and so $R_M = R_{M_n} = R_n$ is a maximal valuation domain.

By the last paragraph R is a locally almost maximal valuation domain. specR = $\{\{0\}\} \cup$ mspecR and $\{v_n\}_{n \in N}$ is a family of valuations of Q with finite character implies R is h-local. By 2.9 R is an almost maximal domain. By 5.2 or the main theorem 9.1, R is an FGC domain. q.e.d.

As an additional comment on this last proof, the R constructed is clearly not a valuation domain, and R is not a P.I.D. since the maximal ideal of $R_{M_n} = R_n$ is not a finitely generated ideal.

Taking a different point of view, the examples 14.1, 14.3, and 14.4 were constructed by finding a field Q and an appropriate family $\{R_i\}_{i \in X}$ of valuation domains satisfying: (i) Q is the quotient field of R_i for all $i \in X$, (ii) R_i is a maximal valuation domain for all $i \in X$, and (iii) $\{R_i\}_{i \in X}$ is an independent set of valuation rings in the sense that no non-zero prime ideal is common to two of the R_i's . In each case the desired R is then $R = \bigcap_{i \in X} R_i$. The divisibility group of R_i is the additive group of rational numbers or the direct sum of n copies of the rational numbers ordered lexicographic. A question one might ask is whether the examples could be simplified by having the group of divisibilities be Z? Surprisingly, the negative answer was known as early as 1933 when in a paper by F.K. Schmidt [27] it was shown that a field cannot have two independent discrete rank one valuation domains (whose quotient fields are the given field) such that both valuation domains are maximal rings. Recall that a valuation domain is discrete rank one if and only if its divisibility group is order isomorphic to Z For a more complete discussion of this topic the reader is referred to [26], page 217, and some recent related topics are included in [30] .

Section 15 Examples of Torch Rings

In this final section we present examples of torch rings.

Example 15.1: (T. Shores and R. Wiegand [28]) There exists a torch ring.

Proof: Let S be an FGC domain which is not local and which is a locally maximal domain. For example S could be the ring of example 14.1, 14.3, or 14.4. Let Q be the quotient field of S. Choose a $J \in \text{mspec}S$. Let $R = S \oplus (Q/S_J)$ where addition of R is componentwise and the multiplication of R is defined by $(s_1,t_1)(s_2,t_2) = (s_1s_2, s_1t_2 + s_2t_1)$ for $s_1,s_2 \in S$ and $t_1,t_2 \in Q/S_J$. We claim that R is a torch ring.

Let $P = \{(0,t) \in R: t \in Q/S_J\}$. Then $P \in \text{spec}R$. Since Q/S_J is a divisible S-module, if $x \in R - P$ and $p \in P$, then $Rx \supset Rp$. If $M \in \text{mspec}S$, then $R(M \oplus \{0\}) = M \oplus \{0\} + P = M \oplus Q/S_J$, and so $\text{mspec}R = \{R(M \oplus \{0\}: M \in \text{mspec}S\}$. In particular R is not local since S is not local.

S is an FGC domain, so S is an h-local, locally almost maximal Bezout domain by 9.4 and 2.9. If $M \in \text{mspec}S - \{J\}$, then $(Q/S_J)_M \cong ((Q/S)_J)_M \cong \{0\}$ by 2.7(2). S_J is a valuation domain by 3.7, and so Q/S_J is a uniserial S_J-module. If $p_1,p_2 \in P$ we can without loss of generality suppose $R_{R(J \oplus \{0\})}p_1 \supset R_{R(J \oplus \{0\})}p_2$. Hence $R_Mp_1 \supset R_Mp_2$ for all $M \in \text{mspec}R$, since $R_Mp_1 \cong R_Mp_2 \cong \{0\}$ for all $M \in \text{mspec}R - \{R(J \oplus \{0\})\}$. By 2.2(1) $Rp_1 \supset Rp_2$, and so P is a non-zero uniserial R-module.

For all $p \in P$ there exists $r \in R - P$ such that $rp = 0$. Hence P is a minimal prime ideal of R. If $r \in R - P$, then $Rr \supset P$, and so P is the unique minimal prime ideal of R. $R/P \cong S$ and S is an h-local domain.

$R_{R(J \oplus \{0\})} \cong S_J \oplus Q/S_J$. S_J is a linearly compact S_J-module since S is a locally maximal ring, and Q/S_J is a linearly comapct S_J-module by 10.10(1) and 1.2(2). Hence $S_J \oplus Q/S_J$ is a linearly compact S_J-module by

1.2(4) , and it follows that $R_{R(J \oplus \{0\})}$ is a maximal ring. If $M \in \text{mspec}S - \{J\}$, then $R_{R(M \oplus \{0\})} \cong S_M$ which is a maximal ring since S is a locally maximal ring. Hence R is a locally almost maximal ring. Since S is a Bezout domain, $x \in R - P$ and $p \in P$ implies $Rx \supset Rp$, and P is a uniserial R-module, it follows that R is a Bezout ring.

We have shown that (i) R is not local, (ii) R has a unique minimal ideal P and P is a non-zero uniserial R-module, (iii) R/P is an h-loca domain, and (iv) R is a locally almost maximal Bezout ring. Hence by definition R is a torch ring. <u>q.e.d.</u>

BIBLIOGRAPHY

1. N. Bourbaki, Éléments de mathématique, Algèbra Commutative, No. 1038, Hermann, Paris, 1964.

2. W. Brandal, Almost maximal integral domains and finitely generated modules, Trans. AMS 183(1973), 203-222. MR 48 # 3956.

3. __________, On h-local integral domains, Trans. AMS 206(1975), 201-212. MR 53 # 10787.

4. __________, Constructing Bezout domains, Rocky Mountain J. 6(1976), 383-399. MR 54 # 2644.

5. __________ and R. Wiegand, Reduced rings whose finitely generated modules decompose, Comm. in Algebra 6(2)(1978), 195-201.

6. J. Dugundji, Topology, Allyn and Bacon, Inc., Boston, 1966.

7. D.T. Gill, Almost maximal valuation rings, J. London Math. Soc. (2) 4 (1971), 140-146. MR 45 # 1904.

8. L. Gillman and M. Jerison, Rings of Continuous Functions, Van Nostrand Co., Princeton, N.J., 1960.

9. R. Gilmer, Multiplicative Ideal Theory, M. Dekker, N.Y., 1972.

10. N. Hindman, On the existance of c-points in $\beta N \setminus N$, Proc. AMS 21(1969), 277-280. MR 39 # 922.

11. Y. Hinohara, Projective modules over semilocal rings, Tohoku Math. J. (2) 14 (1962), 205-211. MR 31 # 4814.

12. M. Hochster, Prime ideal structure in commutative rings, Trans. AMS 142 (1969), 43-60. MR 40 # 4257.

13. I. Kaplansky, Maximal fields with valuations, Duke Math. J. 9(1942), 303-321. MR 3 page 264.

14. __________, Elementary divisors and modules, Trans. AMS 66(1949), 464-491. MR 11 page 155.

15. __________, Modules over Dedekind rings and valuation rings, Trans. AMS 72(1952), 327-340. MR 13 page 719.

16. __________, Infinite Abelian Groups, Univ. of Mich. Press, Ann Arbor, Mich., 1954.

17. W. Krull, Allgemeine Bewertungstheorie, J. Reine Angew. Math. 167(1932), 160-196.

18. J.P. Lafon, Anneaux locaux commutatifs sur lesquels tout module de type fini est somme directe de modules mogenes, J. Algebra 17(1971), 575-591. MR 44 # 202.

19. H. Leptin, Linear kompakte Moduln und Ringe, Math. Z. 62(1955), 241-267. MR 16 page 1085.

20. ________________, Linear kompakte Moduln und Ringe II, Math. Z. 66(1957), 289-327. MR 19 page 245.

21. E. Matlis, Injective modules over Prufer rings, Nagoya Math. J. 15(1959), 57-69. MR 22 # 725.

22. ________________, Cotorsion modules, Mem. AMS No. 49(1964). MR 31 # 2283.

23. ________________, Decomposable modules, Trans. AMS 125(1966), 147-179. MR 34 # 1349.

24. ________________, Rings of type I, J. Algebra 23(1972), 76-87. MR 46 # 5312.

25. R.S. Pierce, Modules over commutative regular rings, Mem. AMS No. 70 (1967). MR 36 # 151.

26. O. Schilling, Valuation Theory, Math. Surveys No. 4, AMS, Providence, R.I., 1950.

27. F.K. Schmidt, Mehrfach perfekte Körper, Math. Ann. 108(1933), 457-472.

28. T. Shores and R. Wiegand, Rings whose finitely generated modules are direct sums of cyclics, J. Algebra 32(1974), 152-172. MR 50 # 4568.

29. A. Tarski, Sur la décomposition des ensembles en sous-ensembles presques disjoints, Fund. Math. 12(1928), 188-205.

30. P. Vamos, Multiply maximally complete fields, J. London Math. Soc. (2) 12(1975), 103-111. MR 52 # 13774.

31. ________________, The decomposition of finitely generated modules and fractionally self-injective rings, J. London Math. Soc. (2) 16(1977), 209-220

32. R.B. Warfield, Jr., Decomposability of finitely presented modules, Proc. AMS 25(1970), 167-172. MR 40 # 7243.

33. S. Warner, Linear compact rings and modules, Math. Ann. 197(1972), 29-43. MR 45 # 6874.

34. R. Wiegand and S. Wiegand, Commutative rings whose finitely generated modules are direct sums of cyclics, Proceedings of the Bicentennial Abelian Group Theory Conference, Springer Lecture Notes in Mathematics, v. 616, 1977, 406 - 423.

35. S. Wiegand, Locally maximal Bezout domains, Proc. AMS 47(1975), 10-14. MR 54 # 5208.

36. ________________, Semilocal domains whose finitely generated modules are direct sums of cyclics, Proc. AMS 50(1975), 73-76. MR 51 # 5587.

37. O. Zariski and P. Samuel, Commutative Algebra, Vol. II, Van Nostrand Co., Princeton, N.J., 1960.

38. D. Zelinsky, Linearly compact modules and rings, Amer. J. Math. 75(1953), 79-90. MR 14 page 532.

Index of Notation and Definitions